Century-Scale Perspective on Water Quality in Selected River Basins of the Conterminous United States

By Edward G. Stets, Valerie J. Kelly, Whitney P. Broussard, III, Thor E. Smith, and Charles G. Crawford

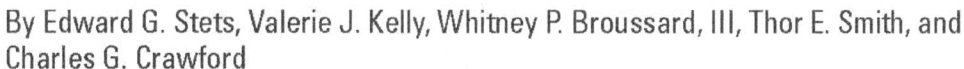

National Water-Quality Assessment Program

Scientific Investigations Report 2012–5225

U.S. Department of the Interior
U.S. Geological Survey

U.S. Department of the Interior
KEN SALAZAR, Secretary

U.S. Geological Survey
Marcia K. McNutt, Director

U.S. Geological Survey, Reston, Virginia: 2012

For more information on the USGS—the Federal source for science about the Earth, its natural and living resources, natural hazards, and the environment, visit http://www.usgs.gov or call 1–888–ASK–USGS.

For an overview of USGS information products, including maps, imagery, and publications, visit http://www.usgs.gov/pubprod

To order this and other USGS information products, visit http://store.usgs.gov

Suggested citation:
Stets, E.G., Kelly, V.J., Broussad, W.P., III, Smith, T.E., and Crawford, C.G., 2012, Century-scale perspective on water quality in selected river basins of the conterminous United States: U.S. Geological Survey Scientific Investigations Report 2012–5225, 108 p.

Contents

Contents—Continued

Figures

Figures—Continued

Tables

Conversion Factors and Datums

Conversion Factors

Inch/Pound to SI

Multiply	By	To obtain
Length		
inch (in.)	2.54	centimeter (cm)
inch (in.)	25.4	millimeter (mm)
foot (ft)	0.3048	meter (m)
mile (mi)	1.609	kilometer (km)
Area		
acre	0.004047	square kilometer (km^2)
square mile (mi^2)	2.590	square kilometer (km^2)
Volume		
million gallons (Mgal)	3,785	cubic meter (m^3)
cubic foot (ft^3)	0.02832	cubic meter (m^3)
acre-foot (acre-ft)	1,233	cubic meter (m^3)
gallon	3.78	liter (L)
Flow rate		
acre-foot per day (acre-ft/d)	0.01427	cubic meter per second (m^3/s)
acre-foot per year (acre-ft/yr)	1,233	cubic meter per year (m^3/yr)
cubic foot per second (ft^3/s)	0.02832	cubic meter per second (m^3/s)
cubic foot per second per square mile [(ft^3/s)/mi^2]	0.01093	cubic meter per second per square kilometer [(m^3/s)/km^2]
gallon per day (gal/d)	0.003785	cubic meter per day (m^3/d)
inch per year (in/yr)	25.4	millimeter per year (mm/yr)
Mass		
ton, short (2,000 lb)	0.9072	megagram (Mg)
ton per day (ton/d)	0.9072	metric ton per day
Hydraulic gradient		
foot per mile (ft/mi)	0.1894	meter per kilometer (m/km)
Application rate		
pounds per acre per year [(lb/acre)/yr]	1.121	kilograms per hectare per year [(kg/ha)/yr]

Conversion Factors and Datums—Continued

Conversion Factors—Continued

SI to Inch/Pound

Multiply	By	To obtain
Length		
millimeter (mm)	0.03937	inch (in.)
centimeter (cm)	0.3937	inch (in.)
meter (m)	3.281	foot (ft)
meter (m)	1.094	yard (yd)
kilometer (km)	0.6214	mile (mi)
Area		
square meter (m^2)	0.0002471	acre
square kilometer (km^2)	247.1	acre
square meter (m^2)	10.76	square foot (ft^2)
square kilometer (km^2)	0.3861	square mile (mi^2)
Volume		
cubic kilometer (km^3)	0.2399	cubic mile (mi^3)
Flow rate		
cubic meter per second (m^3/s)	70.07	acre-foot per day (acre-ft/d)
centimeter per year (cm/yr)	3.937	inch per year (in/yr)
millimeter per year (mm/yr)	0.03937	inch per year (in/yr)
Mass		
gram (g)	0.03527	ounce, avoirdupois (oz)

Temperature in degrees Celsius (°C) may be converted to degrees Fahrenheit (°F) as follows:

$$°F=(1.8×°C)+32.$$

Temperature in degrees Fahrenheit (°F) may be converted to degrees Celsius (°C) as follows:

$$°C=(°F-32)/1.8.$$

Datums

Vertical coordinate information is referenced to the North American Vertical Datum of 1988 (NAVD 88).

Horizontal coordinate information is referenced to the North American Datum of 1983 (NAD 83).

Altitude, as used in this report, refers to distance above the vertical datum.

Century-Scale Perspective on Water Quality in Selected River Basins of the Conterminous United States

By Edward G. Stets, Valerie J. Kelly, Whitney P. Broussard, III, Thor E. Smith, and Charles G. Crawford

Abstract

Nutrient pollution in the form of excess nitrogen and phosphorus inputs is a well-known cause of water-quality degradation that has affected water bodies across the Nation throughout the 20th century. The recognition of excess nutrients as pollution developed later than the recognition of other water-quality problems, such as waterborne illness, industrial pollution, and organic wastes. Nevertheless, long-term analysis of nutrient pollution is fundamental to our understanding of the current magnitude of the problem, as well the origins and the effects. This report describes the century-scale changes in water quality across a range streams in order to place current water-quality concerns in historical context and presents this history on a national scale as well as for selected river reaches. The primary focus is on nutrient pollution, but the development and societal responses to other water-quality problems also are considered. Land use and agriculture in the selected river reaches also are analyzed to consider how these factors may relate to nutrient pollution. Finally, the availability of relevant nutrient and inorganic carbon data are presented for the selected river reaches. Sources of these data included Federal agencies, State-level reports, municipal public works facilities, public health surveys, and sanitary surveys. The availability of these data extends back more than a century for most of the selected river reaches and suggests that there is a tremendous opportunity to document the development of nutrient pollution in these river reaches.

Introduction

Input of excess nutrients in the form of nitrogen and phosphorus is a well-known cause of water-quality degradation that has affected water bodies across the Nation throughout the 20th century. Water effluent from farming operations, municipalities, and industries all contribute to elevated nutrient inputs, and water bodies downstream of these sources are at risk of eutrophication. However, drawing explicit connections between century-long changes in nutrient concentrations and factors believed to be causing those changes requires long-term water-quality and ancillary (for example, land use) data. This report presents the sources of long-term nutrient data for 26 river basins along with brief historical descriptions of events likely to affect water quality within the basins. This study was conducted to evaluate the availability of long-term nutrient concentration data for selected river basins around the country. A strong emphasis was placed on locating data collected by government agencies involved with drinking water quality, biogeochemical investigations, human health and sanitation, as well as resource utilization, such as fisheries management. The availability of ancillary data including streamflow, population density, and land-use characteristics in the river basins of interest also was evaluated in order to put the rivers into context with respect to one another and on a national scale. Comparable water-quality data from many different sources were gathered to document the availability of data for selected river reaches around the country. For most sites, the oldest data identified were collected over a century ago.

Nutrient Pollution

Nutrient pollution is fundamentally caused by excess inputs of nitrogen and phosphorus to water bodies. Humans have dramatically altered the natural cycling of these nutrients. In the case of the nitrogen cycle, humans have added reactive nitrogen to the biosphere through synthetic fertilizer usage, human waste, fossil fuel combustion, and agricultural animal production (Galloway and others, 2004). Phosphorus also is used as a fertilizer and is found in both human and animal waste, mobilized by soil erosion, and contained in many synthetic compounds including some detergents (Edmondson, 1991). Phosphorus is an important component of urban runoff (Novotny, 2003). Additionally, there can be interactions between the phosphorus and nitrogen cycles such that excessive phosphorus inputs stimulate bacterial nitrogen fixation (Schindler, 1977). The global carbon cycle also is experiencing tremendous perturbation through the burning of fossil fuels and changes to biomass production (Le Quéré and others, 2009). Inorganic carbon dissolved in water is not a pollutant, but it can indicate other large-scale anthropogenic disturbances, such as land-use changes (Oh and Raymond, 2006) and hydrologic connectivity, between terrestrial and aquatic environments (Raymond and others, 2008).

Nutrient pollution has resulted in lost biodiversity, eutrophication, and the development of hypoxia in many coastal areas around the world (Diaz, 2001). In the United States, the Gulf of Mexico hypoxic zone stands as perhaps the largest manifestation of anthropogenic nutrient pollution. Agricultural use of synthetic fertilizer is currently a primary driving force of pollution in aquatic ecosystems and the heavily cultivated Mississippi River basin exemplifies these effects. Uneven distribution of nutrient pollution throughout the country reflects both the diversity of agricultural practices (Broussard and Turner, 2009) and differential biogeochemical processing of nutrients in different regions (Schaefer and Alber, 2007).

Several studies successfully used long-term datasets to analyze the relation between changes in land use and stream chemistry. Broussard and Turner (2009) combined modern stream chemistry data, data collected by the U.S. Geological Survey (USGS) in the early 20th century, and the U.S. Department of Agriculture (USDA) Census of Agriculture to explore the correlation between increasing stream nitrate concentrations and agricultural intensity in the Midwestern United States. Robinson and others (2003) compared modern stream chemistry with data collected by the Massachusetts State Department of Health from 1900 to 1914 to explore how stream chemistry has changed with road salting, fertilizer usage, and sulfate aerosol emissions. Goolsby and others (1999) synthesized long-term data from a number of sources including the USGS, the Illinois Environmental Protection Agency, and the Illinois State Water Survey to develop a mass balance model of net nitrogen inputs to the Mississippi River basin. The model strongly implicated anthropogenic fertilizers in total nitrogen concentrations and loads (Goolsby and others, 1999). These analyses and others like them provide a foundation upon which to build additional studies considering century-scale trends in water quality and the land-use factors leading to these changes.

In this publication, the history of water-quality issues since the mid-19th century and the societal response to those issues are briefly discussed, focusing on government actions aimed at improving water access or water quality. The goal of this discussion is to present the evolving nature of water-quality problems and to place current water-quality problems in the context of a long trajectory of change. After discussing the history of national-scale changes in water-quality issues and corresponding management and legislative decisions, methods are summarized for data assembly and management including an algorithm used to synthesize relevant data from disparate collecting agencies. The synthesis of data describing population, land use, and agricultural intensity in the river basins of interest is also summarized. Finally, the individual river reaches of interest and the relevant sources of data are presented for each river reach. In most cases, data exist from several sites close to the primary station for a river reach. The comparability of these sites for nutrient analyses depends on the nutrient being considered, the time frame of the analysis, and the ultimate goal of the analysis. Specific recommendations about which sites could be combined for analytical purposes are not made.

Historical Perspective on Water-Quality Problems

Access to clean and abundant fresh water is fundamental to human welfare and development, but water usage, whether agricultural, industrial, or household, always results in some degree of water-quality degradation. Thus, water-quality issues have become more severe with the expansion of water usage beginning in the mid-19th century (Tarr, 1996). The nature of water-quality concerns and the coincident societal response changed over time. Sometimes the solution to existing water-quality problems spawned new problems, as was the case with the first sanitary sewers that successfully conveyed untreated wastewater away from densely populated areas but resulted in regional waterborne disease outbreaks as the raw waste was delivered to nearby rivers (Tarr, 1996). For the most part, however, new issues arose as societal pressures on water resources increased, such as the emergence of pollution from synthetic organic compounds after World War II (Tarr, 1996). The causes and impacts of water-quality degradation are complex, almost always involving multiple aspects of societal development and multiple classes of stakeholders. In this section, the history of water-quality problems in the United States is briefly discussed, including the responses to these problems. In presenting this history, we hope to emphasize that water-quality issues have changed over time, as has as our understanding of what is needed to successfully address those issues.

Water-Quality Issues in the 19th Century

Population and industrial growth beginning in the 19th century prompted interest in securing adequate water supplies to densely populated areas, primarily located in central and eastern areas of the Nation. The first public waterworks was built in 1802 in Philadelphia to bring water from the Schuylkill River to the city (Tarr, 1996). By 1860, there were 136 public waterworks in the United States and by 1880, they numbered nearly 600 (Tarr, 1996). The availability of tap water allowed affluent citizens to use water much more freely, especially as a means of conveying waste away from households. Substantial increases in per capita water usage and wastewater generation typically occurred after the construction of public waterworks. Most urban areas had a system of cesspools that were designed to capture household and industrial wastewater. When the cesspool was full, the waste was collected and transported to vacant lands, nearby waterways, or sometimes used as fertilizer (Tarr, 1996). These informal systems were quickly overwhelmed after tap water became available and wastewater volumes increased. By the 1870s, there was widespread recognition of the dangers of water-quality degradation from wastewater contamination (Merchant, 2002). Local disease outbreaks and generally unpleasant conditions prompted many cities to contemplate methods for removing wastewater from densely populated areas via sewer systems.

Early sewer systems in the United States were constructed for the sole purpose of transporting wastewater away from cities and into nearby rivers and streams with no consideration given to wastewater treatment. Most large cities used combined sewers designed to carry both stormwater runoff and wastewater, including raw sewage. Sewer construction was accomplished through actions by local water authorities while the Federal government remained strictly concerned with protecting navigation on the Nation's riverways (Scarpino, 1985). Wastewater treatment was considered expensive and unnecessary for many years because running water was believed to be capable of purifying itself (Rauch, 1889). Indeed, studies hailing the ability of heavily polluted streams to rid themselves of waste continued into the 1920s (Hoskins and others, 1927). Nevertheless, the diversion of untreated sewage into streams and rivers had the predictable effect of degrading water quality within the Nation's waterways.

Increases in waterborne diseases during the 1880s and 1890s have been attributed to the relatively primitive sewerage system in the United States (Lewis, 1906; Tarr, 1996). Outbreaks of cholera and typhoid fever increasingly alerted the public to the dangers of discharging untreated wastewater into rivers. For example, recurring typhoid fever and cholera outbreaks in Chicago were linked to contamination of the drinking water source, Lake Michigan (Novotny, 2003). As a response, the Chicago Sanitary and Ship Canal was constructed to reverse the flow of the Chicago River thereby preventing sewage and urban runoff from contaminating the lake; the river was diverted from Lake Michigan to discharge into the Des Plaines River, which drains into the Illinois River, and ultimately the Mississippi River. In other States, public officials began to recognize the need for wastewater treatment and a handful of States responded by creating State public health boards, which were intended to regulate and oversee wastewater discharges into rivers.

Although the initial goal of many State public health boards was to induce wastewater treatment by municipalities, it was quickly discovered that drinking water treatment, primarily through filtration or chlorination, was an effective and much cheaper means of improving drinking water quality and reducing the incidence of waterborne disease (Tarr, 1996). For example, Pennsylvania tasked the public health board with enacting strict guidelines limiting the discharge of untreated waste into rivers following a serious typhoid fever outbreak in Butler, Pa., in 1903. However, when the Pennsylvania Public Health Board required Pittsburgh to produce a plan to treat all of its wastewater in 1910, sanitary engineers working for the city demonstrated that the cost of treating Pittsburgh wastewater would be greater than providing drinking water treatment for 26 downstream cities. The health board refused to proceed and granted Pittsburgh a permit to continue discharging untreated waste into the Ohio River. The permit remained in effect until 1939 (Tarr, 1996). This incident exemplifies the cost-benefit analysis that allowed drinking water treatment to prevail as the preferred method of combating the problem of untreated wastewater discharge into the Nation's rivers.

Modern water quality laws explicitly designate a role for the Federal government in guaranteeing water quality. However, this idea developed very slowly. The Federal role in water issues was essentially limited to navigation and flood control in the 19th and early 20th centuries. By one estimate, more than 100 bills relevant to water-quality standards were introduced to Congress and failed to be enacted before the first, ineffective, national Water Pollution Control Act became law in 1948 (Melosi, 2000). Although the earliest Federal legislation dealing with water quality was the Rivers and Harbors Act of 1899, which declared it unlawful to discharge refuse into navigable waters without a permit (Patrick, 1992), the law narrowly authorized enforcement by the U.S. Army Corps of Engineers (USACE) only if pollution interfered with navigation (Scarpino, 1985). On the Mississippi River, the primary concerns were the removal of snags (fallen trees stuck in the river channel) and ensuring the presence of a navigable channel through the use of wing dams and dredging. Later, flood control on the lower Mississippi River was incorporated into the USACE responsibilities (Barry, 1997). Despite the substantial changes in water quality that were occurring in the late 19th and early 20th centuries, prevailing wisdom at the time dictated that water-quality issues were best decided at the State and local level.

The Water Sanitation Era

Attitudes about Federal responsibilities to safeguard sustainable uses of natural resources, including water, began to change during the reform-minded administration of President Theodore Roosevelt (1901–09). Passage of the Reclamation Act of 1902 permitted a Federal role in the development of water resources in Western States (National Research Council, 2004), and established the Hydrographic Branch of the U.S. Geological Survey, later changed to the Water Resources Branch. In 1908, the Inland Waterways Commission, a Presidential Commission tasked with finding the most appropriate role for the Federal government in guiding water development projects, strongly advocated "multiple-use" projects (Scarpino, 1985). "Multiple-use" development meant that water resources should be managed to provide the greatest overall good, rather than to guarantee the primacy of navigation or flood control. Although the recommendations of the Inland Water Commission were never acted upon, they reflect the evolving perceptions of water as a resource and the appropriate role of the Federal government in managing that resource.

The Public Health Service Act of 1912 specifically directed the Federal U.S. Public Health Service to study "the pollution, direct and indirect, of the navigable waters of the United States" (Cumming and others, 1916). The passage of this legislation was an acknowledgment that water-quality problems in navigable waters were regional in scale and not adequately addressed at the State level. Nevertheless, this bill stopped short of authorizing Federal agencies to enforce pollution controls (Scarpino, 1985). The Public Health Service conducted large-scale studies of pollution in the Ohio River (1914–15), the lower Potomac River (1916), and tidal waters in several Mid-Atlantic States (1916–17). The authors of these studies recognized the negative effects of untreated sewage disposal on water quality, and the link to disease outbreaks. They also began to approach the topic of industrial wastes; although, by at least one account, industrial wastes were considered "not poisonous or otherwise pathogenic" (Cumming and others, 1916), and instead were viewed simply as an impediment to the ability of a river to naturally purify itself (Phelps, 1914).

Concerns over industrial pollution increased as industrial output increased during and after World War I (Merchant, 2002). Demand for petroleum-based fuels and lubricants grew rapidly as internal combustion engines began to replace coal-fired steam engines. Petroleum spills and leaks also increased rapidly and raised public awareness of the nuisance of industrial pollution and the potential to pollute waterways (Scarpino, 1985). This concern coalesced into the Oil Pollution Act of 1924, which authorized the USACE to prohibit vessels from discharging oil into tidal waters and to study the effect oil pollution in rivers may have on navigation and commerce (Scarpino, 1985). This law was the first permitting direct Federal regulation of water pollution, although it was extremely narrow in scope and did nothing to address the problem of water pollution more broadly.

Although drinking water treatment successfully decreased the occurrence of cholera and typhoid fever in large cities during the early part of the 20th century (Melosi, 2000), other negative impacts associated with stream pollution continued to grow. Pollution and overfishing destroyed the inland fisheries in the Illinois River by 1923 (Thompson, 2002). By 1926, the Upper Mississippi River near Minneapolis-St. Paul, Minn., was nearly devoid of fish and was declared a "public health problem" (Twin Cities Metropolitan Council, 2006). In the early 1920s, the American Water Works Association found 248 separate instances of water supplies being harmed by industrial wastes (American Water Resources Association, 1923). The harmful effects of nonpoint source pollution on rivers also began to be recognized. Researchers at the U.S. Fish and Wildlife Service identified pollution as a primary cause for the demise of freshwater mussels throughout the Midwestern United States. Notably, silt eroding from farmland in the Mississippi River basin was believed to be the chief culprit causing the destruction of mussel habitat (Scarpino, 1985). Similarly, the Izaak Walton League recognized the link between nonpoint sources of pollution and water quality.

Their drive to preserve the Upper Mississippi River rested upon the understanding that riparian development would harm fish and wildlife populations and would increase flood risks for downstream communities (Scarpino, 1985). The pressure exerted by the Izaak Walton League prompted the U.S. Congress to approve the creation of the Upper Mississippi River Fish and Wildlife Refuge in 1924 thereby preventing development of river bottomlands between Rock Island, Illinois and Wabasha, Minnesota. However, broad steps to curtail nonpoint source pollution did not occur until after passage of the Federal Water Pollution Control Act (Clean Water Act) in 1972.

River assessments from this time period often noted the presence of high biological oxygen demand (BOD) that arose from inputs of organic material and could result in anoxia, fish kills, and the production of offensive reduced gases (Purdy, 1923; Iowa Division of Public Health Engineering, 1934; Gleeson, 1972). Sanitary sewers and industrial effluent were believed to be the primary cause of elevated BOD in rivers. A landmark study published in 1925 provided a mathematical description of the process of oxygen depletion and re-aeration in rivers. The so-called Streeter-Phelps dissolved oxygen sag curve is still in use; it was developed specifically for point sources of BOD and highlights the emphasis placed on point sources of pollution (Streeter and Phelps, 1925). Low oxygen concentrations in rivers continued to be a problem throughout most of the 20th century. In the most heavily polluted rivers, periods of low oxygen conditions may affect the interpretation of nitrogen concentration data. This is especially relevant if only a single form of nitrogen, such as nitrate, is analyzed. Low oxygen conditions can induce denitrification or can inhibit ammonium oxidation and affect interpretation of these data.

Continued calls to improve navigation and flood control resulted in significant water infrastructure development throughout the country, although these accomplishments sometimes directly contradicted goals to improve river water quality. The Rivers and Harbors Act of 1927 mandated that the Corps of Engineers maintain a 9-foot navigational channel from Cairo, Ill. to St. Louis, Mo. This was accomplished through dredging and construction of wing-dams and so did not disrupt the ability of the river to carry wastewater away from population centers. After passage of the 1930 Flood Control Act, the Army Corps of Engineers began extending the 9-foot navigational channel upstream to Minneapolis-St. Paul, which required that a series of locks and dams be built on the Upper Mississippi River (Scarpino, 1985). The presence of these structures created slack water pools, which did severely curtail the ability of the river to remove municipal wastewater. Lock and Dam 2 in Hastings, Minn., was completed in 1931 and the slack water pool became the final resting place of most of the waste generated by Minneapolis and St. Paul. The resulting conditions eventually prompted the Twin Cities to build the first wastewater treatment plant on the Upper Mississippi (Scarpino, 1985). Construction of this treatment plant began in 1934 and was subsidized by grants from the Public Works Administration, which was created as part of the New Deal (Scarpino, 1985). The willingness of the Federal government to become involved in such projects re-invigorated the stagnating field of wastewater treatment (Melosi, 2000).

Federal spending on water treatment projects increased substantially during the New Deal era. Franklin Roosevelt's Public Works Administration completed nearly 2,500 water projects including wastewater treatment plants (Merchant, 2002). Approximately three-quarters of all projects financed went to towns with populations of 1,000 people or less (Melosi, 2000) and filled an important gap in water infrastructure and sanitation. Small communities began to share in the progress that large cities had made toward combating waterborne disease. National rates of typhoid fever infections decreased from 34 cases per 100,000 people in 1932 to fewer than 4 cases per 100,000 people in 1945 (Merchant, 2002).

During the late 1920s and early 1930s, the United States also was experiencing arguably the worst environmental disaster in its history—the Dust Bowl. Eroding sediment and drought conditions were largely to blame for the severe dust storms experienced throughout semi-arid plains. However, water-driven gully and sheet erosion also were widespread and deposited enormous amounts of sediment into waterbodies throughout the agricultural region (Hansen and Libecap, 2004). At least one study connected serious habitat degradation in rivers and streams with erosion from heavily cultivated areas (Ellis, 1936).

The Rise of Industrial Waste

New Deal investments provided much needed momentum to the goal of improving water quality, but with the outbreak of World War II, emphasis shifted toward promoting industrial output rather than safeguarding water supplies. Improvements made during the New Deal continued to reduce water-quality degradation due to municipal sewage releases (Paavola, 2006), but industrial pollution was significantly increasing. As a result, industrial waste became the largest source of water pollution by the end of World War II (Murphy, 1961). There was growing awareness among public health officials that industrial wastes were becoming an important threat to water supplies. In 1942, the Public Health Service recommended for the first time that water-quality standards include maximum permissible concentrations for industrial waste products having severe physiological effects, such as heavy metals and some synthetic organic compounds (Tarr, 1996). Concern over the quality of waters within the Nation's rivers also increased. During the opening decades of the 20th century, improving water quality was understood to mean decreasing the risk of contracting waterborne illness from consumption of polluted water. With that goal largely accomplished by the early 1940s, more attention began to be paid to improving conditions in the rivers themselves. The U.S. Geological Survey began regularly collecting and publishing water-quality data from streams around the Nation, reflecting the increased emphasis on water quality. However, attempts to improve water-quality standards were largely set aside during World War II because of the need to increase wartime industrial production (Murphy, 1961).

The development of synthetic organic compounds combined with a reluctance to regulate industry during World War II led to a dramatic shift in the magnitude and nature of water pollution following the end of World War II. Synthetic detergents and chlorinated hydrocarbons were widely produced and released into surface waters for the first time and posed a long-term threat to human and ecological health (Tarr, 1996). Groundbreaking work by M.M. Ellis at the newly created U.S. Fish and Wildlife Service demonstrated the dangers of DDT (dichlorodiphyenyltrichloroethane) on aquatic life (Scarpino, 1985). Recognizing the need for strengthened water-quality laws, the U.S. Congress began drafting bills specifying greater involvement of the Federal government in enforcement of water-quality standards. Their work culminated in passage of the 1948 Federal Water Pollution Control Act.

Growth Era

By most measures, the 1948 Water Pollution Control Act was an ineffective piece of legislation that had very little impact on water pollution in the United States, even though it was the first Federal attempt to deal with overall water pollution. The importance of this act stems from the fact that it laid the foundation for a series of improvements ultimately resulting in the 1972 Clean Water Act. As passed in 1948, the Water Pollution Control Act was limited to interstate waters, those that crossed or formed State borders, and authorized the Surgeon General to focus only on pollution that endangered the "health or welfare of persons in a State other than that in which the discharge originated" (Milazzo, 2006). Enforcement of the law was left to the States, and the Surgeon General could only initiate Federal involvement if the polluting State agreed. As a result, in 8 years of existence, not a single pollution abatement order was issued under the 1948 law (Milazzo, 2006).

In 1947, the conversion of a former munitions plant in Muscle Shoals, Ala., to the first large-scale synthetic fertilizer plant also proved to be a pivotal moment for the future of both farming and water quality in the United States. The plant originally used the Haber-Bosch process to produce ammonium nitrate for use in explosives during World War II. After the cessation of hostilities, the demand for explosives decreased, so this material began to be used as fertilizer on farms around the country (Pollan, 2006). Fertilizer application to agricultural fields was not a new phenomenon in 1947, as farmers relied upon animal manure to enrich soil before synthetic fertilizers became widely available. However, fertilizer usage became much more prevalent once synthetic fertilizers were developed in the middle part of the 20th century. Between 1945 and 1951, the amount of fertilizer nitrogen applied to agricultural fields doubled from 0.5 to 1.0 million tons per year and by 1989, it reached 10 million tons per year (Alexander and Smith, 1990; Ruddy and others, 2006); although present in fertilizer in smaller quantities, agricultural applications of phosphorus also were increasing during this time. The acres of cropland on which commercial fertilizer was applied nearly doubled between 1954 and 1978 from 123 to 226 million acres.

During the middle of the 20th century, there was a large-scale shift from agriculture based on small farms relying on organic fertilizers to large and intensively operated industrial farms that focused on growing single crops in monoculture (Novotny, 2003). The advent of an industrial agricultural system highly reliant on synthetic fertilizers paved the way for tremendous increases in yield, and has been implicated in prevalent and widespread nutrient pollution observed in many streams and rivers around the country (Goolsby and others, 1999). Annual yield for corn ranged from 20 to 30 bushels per acre from 1867 to 1944 but began to increase substantially thereafter (National Agricultural Statistics Service, 2010). By 1960, yield reached 50 bushels per acre and then tripled to 150 bushels per acre by 2009 (National Agricultural Statistics Service, 2010). The success in increasing yield occurred through the development of new corn varieties and changes to agricultural practices, including the widespread use of pesticides, but all of it was made possible by the availability of nutrient fertilizers to increase plant growth. Widespread fertilizer application also resulted in elevated nitrogen concentrations in groundwater and surface water and eventually contributed to eutrophication in the Nation's waterways, particularly in coastal areas.

Coincident with the increase in industrialization and specialization in crop production during this time, animal production was developing in a similar direction (MacDonald and McBride, 2009). Livestock increasingly began to be raised at high densities in indoor facilities called concentrated animal feeding operations (CAFOs). CAFOs were introduced during the 1950s to grow poultry and later expanded to include cattle and swine during the 1970s and 1980s. By the turn of the century, CAFOs produced most animals utilized for human consumption in the United States (Burkholder and others, 2007). These concentrations of livestock in a limited area produce excess concentrations of manure-based nutrients that enter streams and rivers from large manure lagoons that may be poorly constructed or that overflow during large precipitation events. Although both nitrogen and phosphorus are present in animal waste, CAFOs are an especially important source for phosphorus (Sharpley and Moyer, 2000; Ruddy and others, 2006).

The understanding of the role of inorganic nutrients, especially nitrogen and phosphorus, as pollutants developed slowly during the second half of the 20th century. Eutrophication of lakes and rivers became increasingly common, but the underlying factors causing these conditions were not widely understood until the 1970s. Phosphorus-containing synthetic detergents are now known to be a particularly acute cause of eutrophication in inland waters, but in 1961, the organizers of a conference about Great Lakes Pollution stated that detergents caused unsightly foam in the water, but were otherwise harmless (De Paul University, 1961). As late as 1969, the offered explanations for eutrophication ranged from increased water temperature to the presence of growth hormones (Schindler, 2006). Richard Vollenweider generally is credited with providing the first scientific explanation of the link between eutrophication of aquatic ecosystems and changes occurring on land, which result in greater nutrient input to waterbodies (Schindler, 2006). Yet his theories continued to be debated well into the 1970s (Schindler, 1977, 2006). Accordingly, water pollution laws did not address nutrient pollution until the 1970s and originally focused only on phosphates arising from wastewater and detergents.

Aside from the emerging recognition of eutrophication as a nutrient issue, general concerns about water supply and water quality also were increasing in the middle of the 20th century. Water usage continued to increase during the 1950s due to rising affluence, placing further demands on aging sewerage and water-treatment facilities. Per capita water usage increased from 137 to 157 gal/d between 1955 and 1965 (Melosi, 2000). Water supply concerns arising from increased usage were further compounded because many treatment facilities had reached or surpassed their design age and capacity, and water sanitation technology had largely stagnated since the 1930s (Melosi, 2000; Milazzo, 2006). However, the response of sanitary engineers to these challenges would lead a top official from the U.S. Public Health Service to retrospectively declare the 1950s the "Decade of Reawakening" for water sanitation (Milazzo, 2006).

The focus of water sanitation and water quality with the original emphasis on preventing waterborne disease, which was largely overcome in previous decades, shifted toward increasing water supply to support postwar economic growth (Milazzo, 2006). The number of people served by public waterworks nearly doubled between 1945 and 1965 from 94 million to 160 million (Melosi, 2000). It was recognized that water pollution contributed to water shortages by decreasing the ability of cities and towns to reuse existing water. People also became increasingly aware of the consequences of water pollution because growing affluence allowed them the leisure time to participate in outdoor activities (Poston, 1961). Updates to the Water Pollution Control Act passed in 1956 provided for greater Federal support to State programs to reduce pollution in interstate, coastal, or Great Lakes waters (Everts and Dahl, 1957). The act preserved State primacy in setting water-pollution standards with the Federal role limited to technical guidance, as well as a Federal grant system to assist in the construction of sewage treatment plants. Federal grants were authorized for projects aiming to improve water quality for the purpose of increasing water supplies for agriculture and industrial use, or "propagation of fish and aquatic life and wildlife" (Everts and Dahl, 1957).

This shift emphasized the emerging thinking that in-stream water quality was a legitimate reason for water treatment. The act also called for substantially increasing the collection of basic chemical and biological water-quality data to document existing conditions.

Despite the renewed interest, however, water quality in streams and lakes continued to deteriorate. Water usage increased 40 percent between 1950 and 1955 from 185 to 262 billion gal/d (Melosi, 2000). Larger water-quality problems accompanied increased water usage, as had been the case since the 1870s. In 1952, the Cuyahoga River caught fire and caused $1.5 million in damage (Melosi, 2000), beach closures due to pollution occurred in every major city on the Great Lakes in 1960 (Chesrow and Hurwitz, 1961), and Lake Erie was recognized as being heavily polluted and eutrophic by the 1960s (Davis, 1961; Melosi, 2000).

Federal involvement in setting water-quality standards was strengthened considerably by passage of the 1965 Water Quality Act. In a departure from earlier legislation, States were required to set water-quality criteria for interstate waters along with plans for implementation and enforcement (Milazzo, 2006). The Secretary of Health, Education, and Welfare oversaw the program and was permitted to set or enforce the criteria if States failed to do so (Melosi, 2000). The act also created the Federal Water Pollution Control Administration, which was tasked with carrying out its objectives. The act attempted to balance the growing needs of water-quality regulation with respect for the tradition of State supremacy in setting water-quality standards. Finding a middle ground between those goals proved difficult, resulting in an ambiguous enforcement environment (Melosi, 2000). The funding of water-quality improvement plans also was insufficient and some of the approved State plans did very little to improve water quality. Therefore, once again, water quality continued to worsen. Yet, this piece of legislation proved to be highly significant as a direct predecessor to the 1972 Clean Water Act.

The magnitude of national water-quality problems at the end of the 1960s was expressed in the Second Annual Report of the President's Council on Environmental Quality: the number of fish killed from pollution reached 41 million in 1969, up from 15 million in 1968 and 6 million in 1960; the shrimp harvest in 1965 was 0.2 percent of what it was in 1936; and, more than 90 percent of drainage basins were considered polluted (Adler and others, 1993). These examples highlight what had become a nearly ubiquitous problem in the United States—water pollution was a serious and rapidly expanding problem.

Government agencies were reorganized in 1970, and the U.S. Environmental Protection Agency (EPA) was established to replace the Federal Water Pollution Control Administration. The EPA was vested with the responsibility to administer and enforce all Federal laws dealing with the environment, including water quality. Shortly after, in 1972, Congress passed the historic amendment to the Federal Water Pollution Control Act, commonly known as the Clean Water Act. This legislation was a substantial departure from the approach taken in the past. Although Federal involvement in water pollution control had been evolving since at least 1948, the burden of proof always laid with enforcement officials to demonstrate that a particular entity was polluting a given body of water and that this pollution was negatively affecting some aspect of the water usage. The stated goal of the Clean Water Act was to "restore and maintain the chemical, physical and biological integrity of the Nation's waters" (Adler and others, 1993). Therefore, the logical starting point was that the Nation's waters ought to be free from pollution and entities wishing to discharge waste into waterways needed to provide justification and receive permission to do so.

The Clean Water Act, therefore, reversed the previous approach by requiring dischargers to obtain a permit via the National Pollution Discharge Elimination System (NPDES), which defines the allowable limit for the pollution load they could discharge into a waterway. Implicit in this approach is a distinction between point sources of pollution, defined as discharges, whose sources can be clearly identified and controlled, and nonpoint or diffuse sources. This second category was defined in the Clean Water Act as a range of pollution sources primarily from runoff and atmospheric deposition, or sources that cannot be precisely identified and are difficult to measure and quantify (Novotny, 2003). Delivery of pollutants from nonpoint sources is often sporadic, linked to events such as storms that generate runoff from agricultural fields or urban areas. Therefore, these sources are more difficult to control than discharges from traditional point sources.

Recent History

Following the requirement in the 1972 Clean Water Act that dischargers of pollution must abide by limits established by the NPDES permit, treatment of effluent wastewater was greatly enhanced. The Act specified that municipal sewage treatment facilities provide the best treatment practicable or available by the late 1970s and early 1980s, with a minimum of secondary treatment that includes on-site biological oxidation of organic wastes in wastewater (Patrick, 1992). States were directed to define water-quality standards for all State waters, and prepare water-quality management plans for achieving those standards based on establishing relationships between water quality and land use. The plans were mandated to quantify total maximum daily loads (TMDLs) for pollutants in streams that did not meet the standards, defining the permitted load for each target constituent from all contributing sources, including point and nonpoint sources. Nonpoint sources generally were assumed to encompass all agricultural operations, including runoff from fields, irrigation return flows, and stormwater discharges from CAFOs.

The primary emphasis in the Clean Water Act was municipal and industrial point sources, whose discharges were limited and controlled by the NPDES permits. No mechanisms for control of nonpoint sources were implemented by the Act, and no incentives for complying with pollution abatement programs were specified for these sources. Nonetheless, the enactment of a land-use planning process in the development of management plans represented the first formal recognition that control of point sources would not be sufficient to solve the full range of pollution problems in the country (Patrick, 1992).

During subsequent years, the Clean Water Act was amended. In 1974, health- and nuisance-related standards for drinking water were established. In 1977, amendments added the Rural Clean Water Program to support implementation of best management practices controlling pollution from nonpoint sources (Patrick, 1992). More emphasis was placed on developing measures to control agriculturally related pollution, with guidance provided by the Department of Agriculture. Gradually, legislative focus began to incorporate explicit land use changes in the form of wetland protection with the passage of the Wetlands Resources Act in 1986, which provided Federal funding for acquisition and conservation of wetlands. In 1987, further amendments to the Clean Water Act required States to develop more stringent management programs specifically targeting nonpoint sources, including agricultural sources. Still, legislation was largely ineffective because of insufficient funding and limited consequences of noncompliance. A significant component of the dilemma relative to controlling nonpoint source pollution lay in the strong relationship between land use and nonpoint source discharges, because management of land use was not a Federal concern but rather within the legislative domain of State and local government (Patrick, 1992).

Over the course of the last two decades of the 20th century, a number of important research programs were implemented to investigate the association between land use and nonpoint pollution. The Soil and Water Resources Act of 1977 mandates that the Secretary of Agriculture conduct studies via the Soil Conservation Service to assess national soil and water conditions, and to develop conservation programs in partnership with the States. The USDA initiated the Rural Clean Water Program in 1980, which provides funding for long-term watershed projects focused on identifying successful approaches to controlling nonpoint pollution from agricultural sources (Novotny, 2003). The National Urban Runoff Project was carried out in the 1980s to study the characteristics of urban runoff; results laid the foundation for the expansion of the NPDES permit coverage in the mid-1990s to include runoff from urban and industrial sources (Novotny, 2003). Finally, in 1983, the States of Pennsylvania, Maryland, and Virginia, as well as the mayor of the District of Columbia and the administrator of the EPA, established the Chesapeake Bay Agreement to investigate the causes of declining water quality in Chesapeake Bay.

Excess nutrients from upstream sources were found to be a significant factor. The drainage basin model produced by this collaboration provided the means to articulate the relative contribution of point and nonpoint sources, and especially to identify specific areas associated with elevated nutrient loads (Novotny, 2005). This program is ongoing and among the forefront in the Nation for developing management programs to assist farmers in reducing soil and nutrient losses.

Despite these efforts, more than 50 percent of receiving water bodies were not meeting the water-quality goals established by the Clean Water Act by the end of the 20th century (Novotny, 2003). As a result of this lack of comprehensive progress, dozens of lawsuits were filed in the mid-1990s against the EPA and the States by environmental groups to implement the TMDL requirement of the Clean Water Act. Although subject to legal challenge, the TMDL process is currently recognized as the most effective legal tool that can result in minimizing pollution loads from nonpoint sources, which are widely recognized to be primarily agricultural and urban sources (Novotny, 2003).

In the last decade of the 20th century, new water-quality concerns arose about the effect of hydraulic fracturing (fracking) procedures utilized in commercial natural gas production from shale formations (Curtis, 2002). Abundant natural gas resources have been identified in shale basins that include the Barnett Shale, New Albany Shale, Lewis Shale, and Marcellus Shale, potentially affecting a number of rivers in our study (Kargbo and others, 2010). Fracking procedures include injecting water horizontally through deep well casings at sufficient pressure to create fractures in shale-gas formations so that the gas can be extracted. This technique requires large volumes of fresh water augmented by drilling additives. The wastewater contains chemicals found in the geological formation as well as those found in the additives (Kargbo and others, 2010). Contaminants associated with fracking include ammonium and other salts, toxic metals, a wide range of organic compounds, and radionuclides (Kargbo and others, 2010). At the time this report was written, effective treatment or regulation of fracking wastewater was not well developed, although the EPA was beginning to develop national standards for fracking wastewater discharged to natural streams (Kusnetz, 2011).

This study is focused on compiling data from selected river basins that will be useful to evaluate how these long-term changes in policy have coincided with changes in nutrient concentrations. The emphasis is on acquiring data to support assessment of changes throughout the entire course of the 20th century, to the greatest degree possible. This perspective is unprecedented and challenging to develop, especially given the changes in sampling and analytical methodologies over such a lengthy period. Nonetheless, it is a valuable and critical one for understanding the impact and effectiveness of legislation in controlling excess delivery of nutrients to streams and rivers of the Nation.

Methods

Stream-Gaging Station Selection

Sites were selected based on the availability of data and for representation of different basin types in the study. Therefore, basins were included to represent a high degree of agricultural development, some with urban development, some with very little overall development, and sites which represented different regions of the conterminous United States. In all, 26 individual stations were selected along rivers reaches of interest (table 1, fig. 1). In some cases, several stations were selected within the same river basin with the result that some basins are nested within other larger basins. River basins with multiple stations nested within them include the Mississippi, the Illinois, the Ohio, and the Upper Colorado (table 1, fig. 1). For the purposes of describing water-quality history and ancillary data trends, the river basin names as they appear in table 1 are used. Additionally, river basins and associated USGS stream-gaging stations are shown in table 1 and described in greater detail in individual river basin sections. In order to place the river basins into greater spatial context, we describe them in terms of physiographic provinces (fig. 2).

Data Sources

An initial assessment of data availability was performed by aggregating all data available in the USGS National Water Information System (NWIS; http://waterdata.usgs.gov/nwis) and the EPA Storage and Retrieval System (STORET; http://www.epa.gov/storet). Data outside of these databases were found by searching the USGS Publications Warehouse, State agency publications, water-treatment facilities, and Canadian Federal agencies, and U.S. Federal agencies outside of the USGS and EPA. The availability of data for each site described in this publication reflects the ultimate outcome of these searches. Streamflow, nutrient, and inorganic carbon data availability are reported, but often major ions, measures of oxygen demand, and other field parameters, such as turbidity, residue on evaporation, and water temperature are available as well.

Table 1.   River basins included in this study and their associated reference stream gaging station.

[**Abbreviations:** No., number; USGS, U.S. Geological Survey]

Geographic region	River basin	Station name	USGS station No.	Latitude	Longitude
Eastern	Connecticut	Connecticut River at Thompsonville, Conn.	01184000	41°59'14"	72°36'19"
	Delaware	Delaware River at Trenton, N.J.	01463500	40°13'18"	74°46'41"
	Schuylkill	Schuylkill River at Philadelphia, Pa.	01474500	39°58'04"	75°11'19"
	Potomac	Potomac River (adjusted) near Washington, D.C.	01646502	38°56'58"	77°07'39"
	James	James River at Cartersville, Va.	02035000	37°40'16"	78°05'09"
Great Lakes	Maumee	Maumee River at Waterville, Ohio	04193500	41°30'00"	83°42'46"
	St. Lawrence	St. Lawrence River at Cornwall Ontario, near Massena, N.Y.	04264331	45°00'22"	74°47'42"
Central	Middle Ohio	Ohio River at Louisville, Ky.	03294500	38°16'49"	85°47'57"
	Lower Ohio	Ohio River at Cairo, Ill.	03614500	36°59'45"	89°08'36"
	Des Moines	Des Moines River at Keosauqua, Iowa	05490500	40°43'40"	91°57'35"
	Middle Illinois	Illinois River at Kingston Mines, Ill.	05568500	40°33'11"	89°46'38"
	Lower Illinois	Illinois River at Valley City, Ill.	05586100	39°42'12"	90°38'43"
	Upper Mississippi	Mississippi River at Grafton, Ill.	05587450	38°58'05"	90°25'44"
	Missouri	Missouri River at Hermann, Mo.	06934500	38°42'35"	91°26'19"
	Middle Mississippi	Mississippi River at Thebes, Ill.	07022000	37°12'59"	89°28'03"
	Arkansas	Arkansas River at Murray Dam, near Little Rock, Ark.	07263450	34°47'35"	92°21'30"
	Lower Mississippi	Mississippi River at Baton Rouge, La.	07374000	30°26'44"	91°11'30"
	Brazos	Brazos River at Richmond, Tex.	08114000	29°34'57"	95°45'28"
Upper Colorado River	Gunnison	Gunnison River near Grand Junction, Colo.	09152500	38°58'60"	108°27'02"
	Colorado at Cisco	Colorado River near Cisco, Utah	09180500	38°48'38"	109°17'36"
	Green	Green River at Green River, Utah	09315000	38°59'10"	110°09'04"
	San Juan	San Juan River near Bluff, Utah	09379500	37°08'49"	109°51'53"
	Colorado at Lee's Ferry	Colorado River at Lees Ferry, Ariz.	09380000	36°51'53"	111°35'18"
Western	Santa Ana	Santa Ana River below Prado Dam, Calif.	11074000	33°53'00"	117°38'43"
	San Joaquin	San Joaquin River near Vernalis, Calif.	11303500	37°40'34"	121°15'59"
	Willamette	Willamette River at Salem, Oreg.	14191000	44°56'39"	123°02'34"

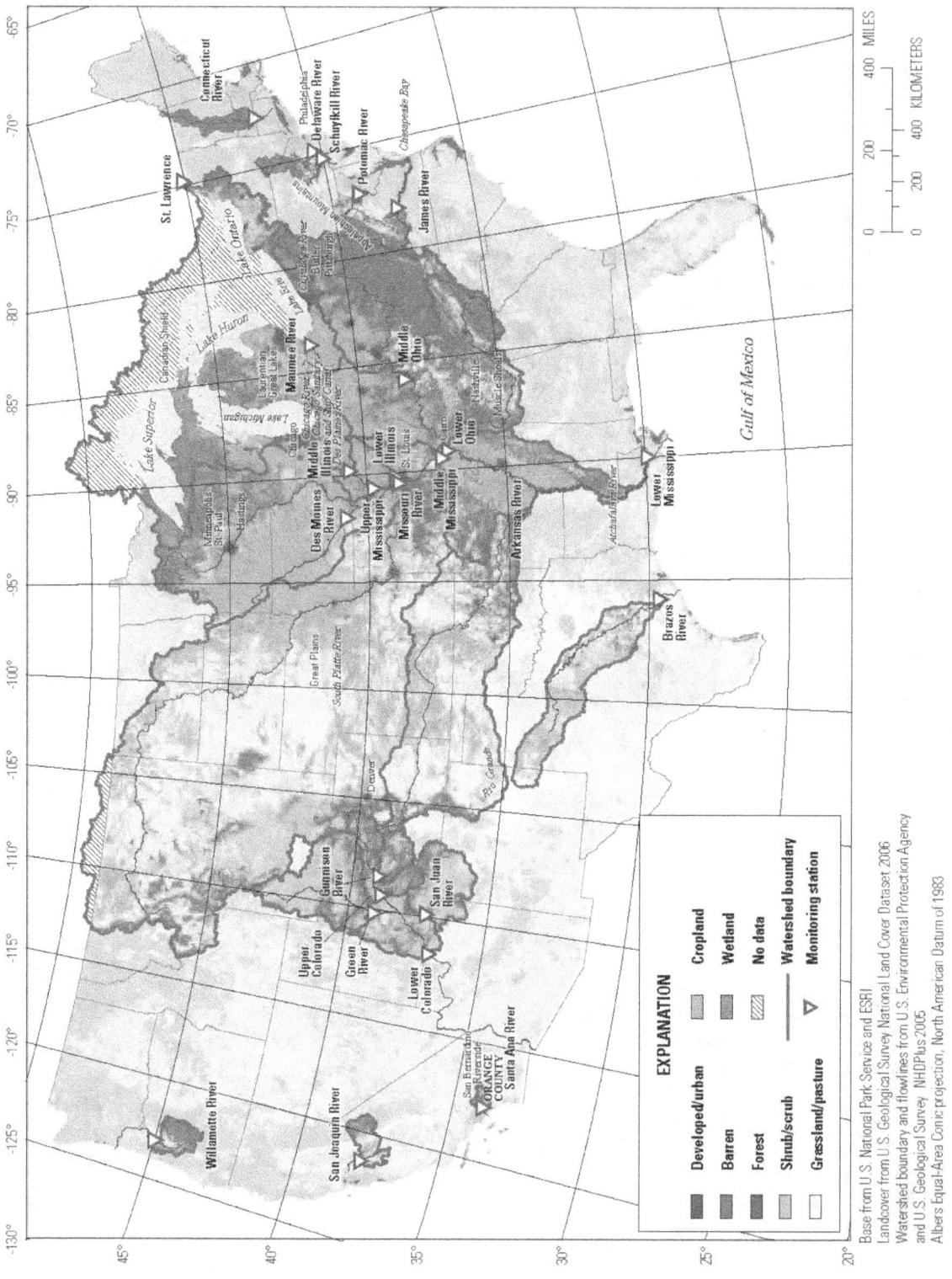

Base from U.S. National Park Service and ESRI
Landcover from U.S. Geological Survey National Land Cover Dataset 2006
Watershed boundary and flowlines from U.S. Environmental Protection Agency
and U.S. Geological Survey NHDPlus 2005
Albers Equal-Area Conic projection, North American Datum of 1983

Figure 1. Land-cover data and river basins included in this study. Monitoring stations are the reference stream gaging stations for all river reaches of interest (see table 1).

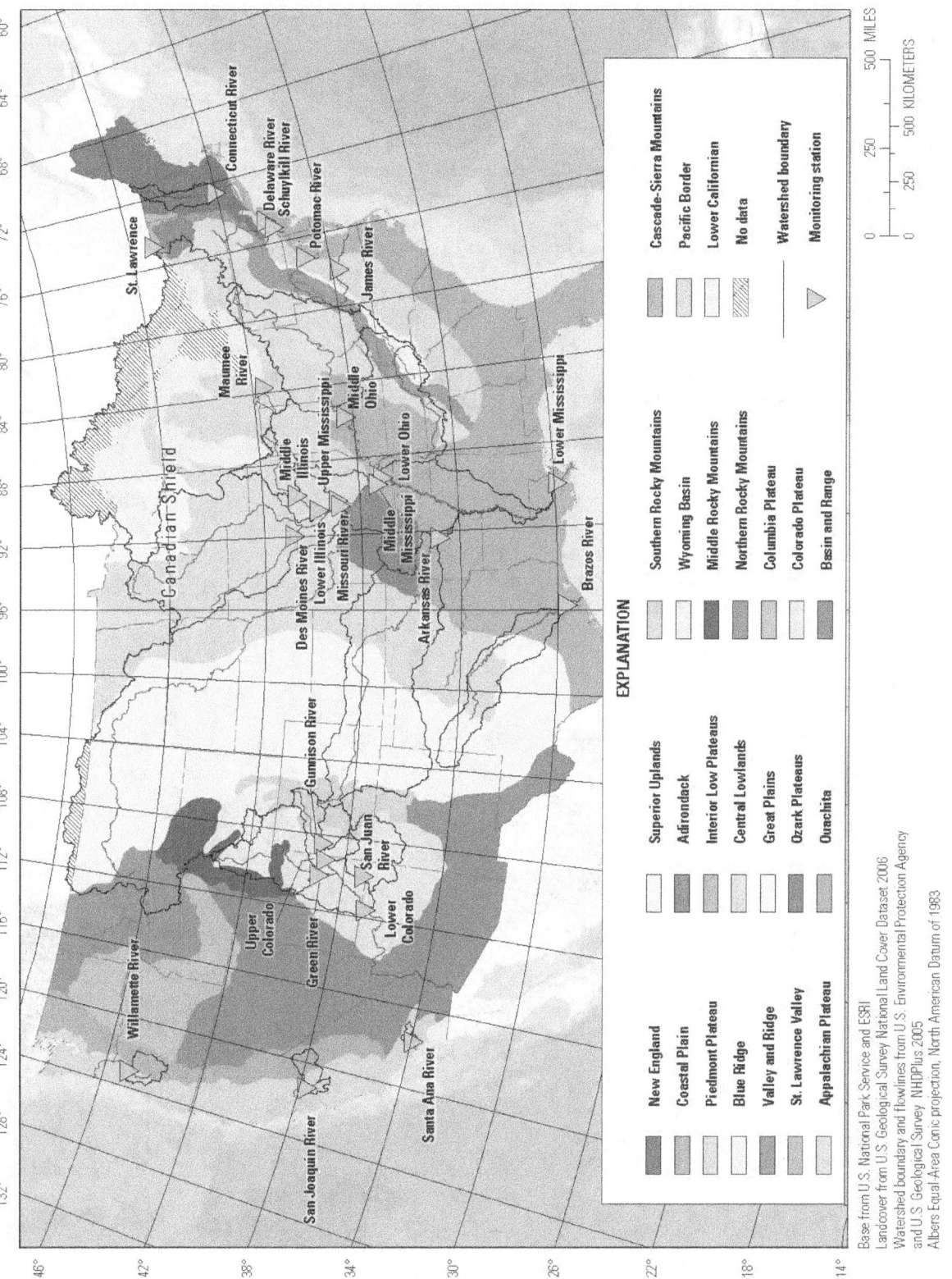

Figure 2. Physiographic provinces of the conterminous United States along with river basins in this study. Monitoring stations are the reference stream gaging stations for all river reaches of interest (see table 1).

Base from U.S. National Park Service and ESRI
Landcover from U.S. Geological Survey National Land Cover Dataset 2006
Watershed boundary and flowlines from U.S. Environmental Protection Agency
and U.S. Geological Survey NHDPlus 2005
Albers Equal Area Conic projection, North American Datum of 1983

Parameter Codes Used in this Study

Nitrogen and Phosphorus Codes

In our analysis of the availability of nitrogen data, several nitrogen species were considered including nitrate, ammonium, organic nitrogen, and total nitrogen. For phosphorus, both dissolved and total phosphorus were considered. Where necessary, concentrations were transformed so that all were reported consistently at the elemental level (for example, ammonia was transformed to N). Algorithms to calculate missing nutrient species were utilized when possible and the calculated values were considered to be equivalent to the reported values. Nitrate plus nitrite is reported as nitrate if the latter was missing. Similarly, filtered and unfiltered nitrate measurements are considered equivalent because nitrate is typically a small portion of the particulate pool (Harrington and Harrington, 2009). Table 2 lists the parameter codes used in the algorithms to identify and calculate (where necessary) the range of nitrogen and phosphorus species in this analysis.

Inorganic Carbon Codes

At least 32 separate parameter codes express some form of inorganic carbon in NWIS and STORET. In order to make these parameter codes comparable, dissolved inorganic carbon was calculated, in milligrams carbon per liter (mg C/L), for any sample reporting dissolved inorganic carbon, alkalinity, acid-neutralizing capacity, or bicarbonate. In some cases, a sample listed several parameters relating to dissolved inorganic carbon. Parameters were preferentially used that would produce the best estimate of dissolved inorganic carbon in a given sample. Filtered, field-titrated measurements were given highest priority because significant changes to alkalinity and acid neutralizing capacity (ANC) can occur in the holding time between field and laboratory observations. Reported alkalinity or ANC values were preferred over reported HCO_3 values because HCO_3 was determined in older samples by titration to pH 4.5, which has been shown to be dependent on temperature and ionic strength of the water sample (Barnes, 1964). All alkalinity parameter codes, their priority of selection, and the frequency of occurrence in our database are listed in table 3. Parameter code 00410 comprised 65 percent of all alkalinity observations in our database. Prior to 1980, measurements of unfiltered water titrated to pH 4.5 were reported as parameter code 00410 regardless of whether the titration was performed in the field or the laboratory. However, it is assumed that most of these samples were analyzed in a laboratory setting (U.S. Geological Survey, 1988). In the USGS database, roughly 80 percent of the observations

having parameter code 00410 occur prior to 1980. Subsequent to 1980, parameter code 00410 was reserved for analyses performed in the field and parameter code 90410 was used to denote the same analysis performed in a laboratory. Roughly 7 percent of all of the alkalinity observations originate from parameter code 90410. Parameter code 00440 (bicarbonate), determined by fixed endpoint titration on unfiltered water samples in the field, comprised approximately 10 percent of the database. This parameter code was used most frequently between 1950 and 1970. Analyses of alkalinity by incremental titration performed in the field, parameter code 39086, also comprised roughly 10 percent of the observations. All other parameters were less common, none of which made up more than 2 percent of the total number of observations.

Alkalinity and ANC are expressed in terms of the amount of inorganic carbon available to react with divalent calcium (Ca^{2+}), given in the units mg $CaCO_3$ L^{-1}. Converting to mg C L^{-1} requires both stoichiometric conversion and consideration of the charge on the inorganic carbon molecule, because twice as much carbon is available to react with Ca^{2+} if it exists as HCO_3^- than if it exists as CO_3^{2-}. In order to calculate the distribution of inorganic carbon species, dissociation constants Ka_1 and Ka_2 were calculated as functions of temperature (Plummer and Busenberg, 1982) and the proportionate distribution of CO_3^{2-} and HCO_3^- (rc) as

$$r_C = \frac{1}{\left(1 + \dfrac{[H^+]}{K_2}\right)}, \quad (1)$$

where $[H^+]$ is the hydrogen ion concentration, calculated as 10^{-pH}. Individual constituents of dissolved inorganic carbon were then calculated as:

$$HCO_3^- \text{ mg C } L^{-1} = (1 - r_C) \times 0.24 \times (\text{ALK mg } CaCO_3 \text{ } L^{-1}) \quad (2)$$

$$CO_3^{2-} \text{ mg C } L^{-1} = r_C \times 0.12 \times (\text{ALK mg } CaCO_3 \text{ } L^{-1}) \quad (3)$$

$$CO_2 \text{ mg } L^{-1} = \frac{(HCO_3^-)(H^+)}{12 \times 1000 \times Ka_1} \quad (4)$$

where ALK is the reported alkalinity or ANC, and HCO_3^- in equation (4) was calculated as in equation (1) and expressed in mg C L^{-1}. Dissolved inorganic carbon was then calculated as the sum of CO_2, HCO_3^-, and CO_3^{2-}.

Table 2. Parameter codes used in nitrogen and phosphorus calculations.

Constituent	Abbreviation	Parameter code	Parameter description
Organic N	TON	00605	Organic nitrogen, water, unfiltered, milligrams per liter
Ammonia	NH_3	00610	Ammonia, water, unfiltered, milligrams per liter as nitrogen
		71845	Ammonia, water, unfiltered, milligrams per liter as NH_4
		71846	Ammonia, water, filtered, milligrams per liter as NH_4
		00608	Ammonia, water, filtered, milligrams per liter as nitrogen
NH_3 + Organic N	TON + NH_3	00625	Ammonia plus organic nitrogen, water, unfiltered, milligrams per liter as nitrogen
		00635	Ammonia plus organic nitrogen, water, unfiltered, milligrams per liter as nitrogen
Nitrite	NO_2	00615	Nitrite, water, unfiltered, milligrams per liter as nitrogen
		71855	Nitrite, water, unfiltered, milligrams per liter
		71856	Nitrite, water, filtered, milligrams per liter
Nitrate	NO_3	00620	Nitrate, water, unfiltered, milligrams per liter as nitrogen
		71850	Nitrate, water, unfiltered, milligrams per liter
		00618	Nitrate, water, filtered, milligrams per liter as nitrogen
		71851	Nitrate, water, filtered, milligrams per liter
Nitrate + Nitrite	[1]NO_3	00630	Nitrate plus nitrite, water, unfiltered, milligrams per liter as nitrogen
		00631	Nitrate plus nitrite, water, filtered, milligrams per liter as nitrogen
Total nitrogen	TN	00600	Total nitrogen, water, unfiltered, milligrams per liter
		62855	Total nitrogen (nitrate + nitrite + ammonia + organic-N), water, unfiltered, analytically determined, milligrams per liter
		71887	Total nitrogen, water, unfiltered, milligrams per liter as nitrate
Dissolved phosphorus	TDP	00666	Phosphorus, water, filtered, milligrams per liter as phosphorus
		71888	Phosphorus, water, filtered, milligrams per liter as phosphate
		00671	Orthophosphate, water, filtered, milligrams per liter as phosphorus
		00660	Orthophosphate, water, filtered, milligrams per liter
		91004	Orthophosphate, water, filtered, micrograms per liter as phosphorus
		99122	Orthophosphate, water, filtered, field, milligrams per liter
		99893	Phosphorus, water, filtered, modified jirka method, milligrams per liter
Total phosphorus	TP	00665	Phosphorus, water, unfiltered, milligrams per liter as phosphorus
		71886	Phosphorus, water, unfiltered, milligrams per liter as phosphate

[1] NO_3 and NO_2 + NO_3 are considered equivalent in historical data because NO_2 concentration is typically much lower than NO_3 concentration.

Table 3. Alkalinity parameter codes shown in descending hierarchy along with their frequency in the database used in this study.

[Symbol: <, less than]

Constituent	Parameter code	Frequency	Percentage of frequency
Dissolved inorganic carbon (DIC)	00691	2,273	0.56
Alkalinity – Filtered, field	29802	934	0.23
	39086	40,084	9.85
	00418	204	0.05
	39036	6,580	1.62
Bicarbonate – Filtered, field	63786	0	0
	00453	663	0.16
	29804	1	<0.01
Acid-neutralizing capacity – Unfiltered, field	29813	0	0
	00419	3,944	0.97
	00410	264,963	65.14
	00416	7	<0.01
	00417	3,710	0.91
	90410	28,603	7.03
	95410	56	0.01
	00413	56	0.01
	00431	3,058	0.75
	00685	482	0.12
Bicarbonate – Unfiltered, field	00450	1	<0.01
	99440	50	0.01
	00440	41,903	10.3
	00449	0	0
	90440	27	0.01
	00451	5,886	1.45
	95440	223	0.05
Alkalinity – Filtered, laboratory	29803	669	0.16
	39087	56	0.01
	00421	0	0
	29801	1,641	0.4
Bicarbonate – Filtered, laboratory	29806	0	0
Acid-neutralizing capacity – Unfiltered, setting not specified	00431	0	0
	00425	688	0.17

Stream Discharge and Runoff Calculations

All discharge and runoff values presented in this publication were derived from information obtained from NWIS. Annual discharges were calculated from the annual average discharge reported in NWIS and averaged over at least 10 years. In most cases, annual discharge was averaged over much longer time periods. For drainage basins in which discharge was subject to significant hydrologic modification, water withdrawals, or water infrastructure development, annual averages are presented from after the modification. Therefore, the values presented should reflect modern annual discharge unless otherwise noted.

Runoff was calculated as annual discharge divided by drainage area and expressed as millimeters per year and used as a general descriptive characteristic of water availability in the basins, as well as a way of comparing the river reaches of interest to one another. In some cases, drainage basins contain significant non-contributing areas, such as endorheic basins. For these basins, in NWIS, the contributing drainage area is sometimes distinguished from total drainage area. In these cases, we calculated runoff from total drainage area but discuss how runoff may differ if only contributing area was used.

Ancillary Data Methodology

We analyzed ancillary data that are presumed to be relevant to understanding the changes in water quality associated with each of the basins of interest. In this sense, ancillary data were considered to represent potential causative factors of observed water chemistry in these river basins. The ancillary data presented in the following sections are not meant to be exhaustive, but rather allow us to place river basins in context historically and with respect to one another. These potential causative factors include indicators of agricultural intensity, fertilizer usage, nitrogen and phosphorus inputs from livestock, and population density.

In order to investigate changes in land use associated with agriculture, county-level data were evaluated from the Census of Agriculture beginning in 1870. Census of Agriculture data from 1870 to 1910 are from public document ICPSR 2896 (Haines and Inter-university Consortium for Political and Social Research, 2004). Agricultural data from 1930 and 1940 were derived from the decennial population census conducted in those years. For census years 1920, 1949, and all years from 1954 to 2002, Census of Agriculture data tables were digitized and organized into data files (M. Haines, University of Colgate, written commun., 2004). Fertilizer data were used from two USGS publications, the first reported annual county-level data from 1945 to 1985 (Alexander and Smith, 1990) and the second reported county-level fertilizer usage from 1987 to 2001 (Ruddy and others, 2006). No fertilizer usage data were available for 1986 and so this year was excluded from our analysis.

In each case, nitrogen and phosphorus inputs were calculated from both farm and non-farm fertilizer usage. Nitrogen and phosphorus inputs were also estimated from livestock and farm animals using Census of Agriculture data and following the general methods explained in Ruddy and others (2006). County-level population data were downloaded from the National Historical Geographic Information System (Minnesota Population Center, 2011). Population census data are available on a decadal basis, and Census of Agriculture data are available on a decadal basis from 1870 to 1920 with subsequent agricultural census years more frequent, every 4 to 5 years, up to 2002. State-level data from the USDA National Agricultural Statistical Service annual survey were also used to estimate corn harvest on an annual basis. County-level corn harvest data from the closest agricultural census year were used to estimate the proportion each county contributed to state-level corn harvest. State-level survey data were used to estimate county-level corn harvest in the non-census years.

Animal manure is a nonpoint source of nitrogen and phosphorus so the production of nitrogen and phosphorus also was calculated in the basins of interest. To calculate animal livestock contributions to nitrogen and phosphorus inputs, we tabulated the population of several animal groups, calculated the production of nutrients by each animal group, and then summed the contribution of each animal group to calculate total animal contribution of nitrogen and phosphorus. The animal groups consisted of cattle, horses, sheep, chickens, and turkeys. These groupings and animal nutrient production rates were identical to those presented in Ruddy and others (2006), except that all cattle were included in a single group rather than being divided into sub-categories. Animal contributions to nitrogen and phosphorus production were calculated for the agricultural census years and then interpolated on an annual time scale to allow direct comparison with annual fertilizer input data.

Agricultural and population statistics on a basin scale were calculated by spatially relating county-level data to basin geography. The historical U.S. County boundary file was used to create lists of counties contained or partially contained in each river basin at 10-year intervals beginning in 1870 and continuing through 2010.

The contribution of each county to basin characteristics was tabulated on an areal basis such that the metrics reported for any county were weighted by the areal proportion of the county residing within the river basin. For example, if a county lies entirely within a river basin, all of the people and farmland of that county were considered to contribute to the river basin total. For counties partially inside a basin, the county totals were multiplied by the areal proportion of the county inside the basin. This approach has the limitation of assuming that farmland and population are all distributed evenly throughout a county, which is often a poor assumption. For large basins encompassing many counties, the assumption probably has minimal effect on our calculations. However, for smaller drainage basins, significant artifacts can be introduced. One extreme example is the Santa Ana River basin, which is situated in small, densely populated areas of San Bernardino, Riverside, and Orange Counties in southern California. Riverside and San Bernardino Counties are large with population centers clustered in the Santa Ana River basin. The result is that when calculated on a county-wide basis, the population density of the Santa Ana River basin appears to be small. Because this river basin was such an outlier, a different approach was used to calculate population density. We used census tract data, which is presented at a finer spatial scale, from 1950 and 2000 to determine the proportion of the county population residing within the river basin and used this coefficient to calculate population density for each of the population census years. Population densities calculated for other river basins were reasonably close to values published elsewhere (Benke and Cushing, 2005), so no further corrections were made.

For basins located partially within Canada, including the Missouri, Mississippi, Connecticut, and Saint Lawrence, the Canadian portion of the basin was excluded from our calculation of ancillary basin characteristics. For the Missouri, Mississippi, and Connecticut River basins, this had a minimal impact on our analysis because only a very small fraction of the drainage basin is located within Canada (1.4, 0.7, and 1.2 percent, respectively). However, more than 40 percent of the Saint Lawrence River basin is located within Canada so it must be assumed that basin characteristics are similar in Canada and the United States. This issue is addressed in greater detail in section, "Saint Lawrence River."

Recent Ancillary Data for all Basins

Population

According to the 2000 population census, the most densely populated basin in this study was the Santa Ana River, with almost 330 people/km² (U.S. Census Bureau, 2000) (fig. 3). The Schuylkill River basin, which includes parts of Philadelphia and western suburbs, also had more than 300 people/km²; and the Illinois River, which includes densely populated areas of northern and central Illinois, had a population density of 150 and 240 people/km² in the lower and middle sections, respectively (fig. 3). By contrast, the five stations considered in the Colorado River basin had the lowest population densities, ranging from approximately 2 to 6 people/km² (fig. 3). Overall, approximately one-half of the river basins included in our study had population densities greater than the national average in the year 2000 of 31 people/km² (fig. 3).

Farmland Area

Figure 4 shows the percentage of each river basin with land usage specified as farmland, according to the 2002 Census of Agriculture (U.S. Department of Agriculture, 2005). Farmland was separated into two categories: cropland and non-cropland (which includes rangeland and other uses).

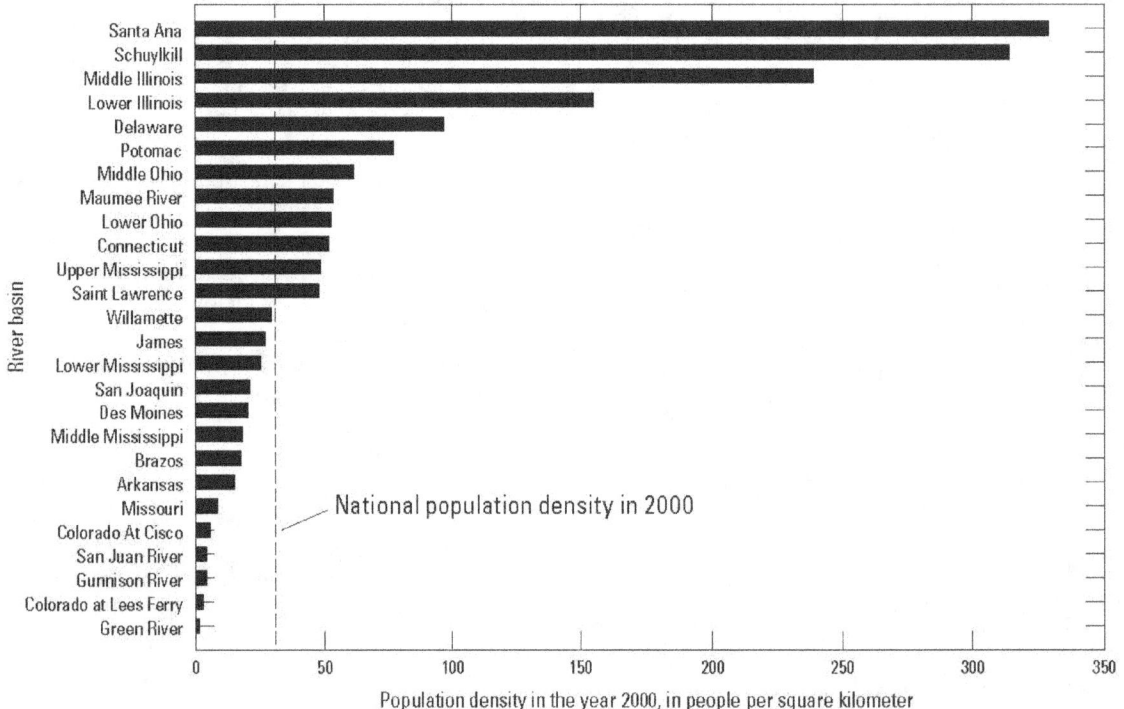

Figure 3.   Population density from 2000 census (U.S. Census Bureau, 2000) for all river basins included in this study.

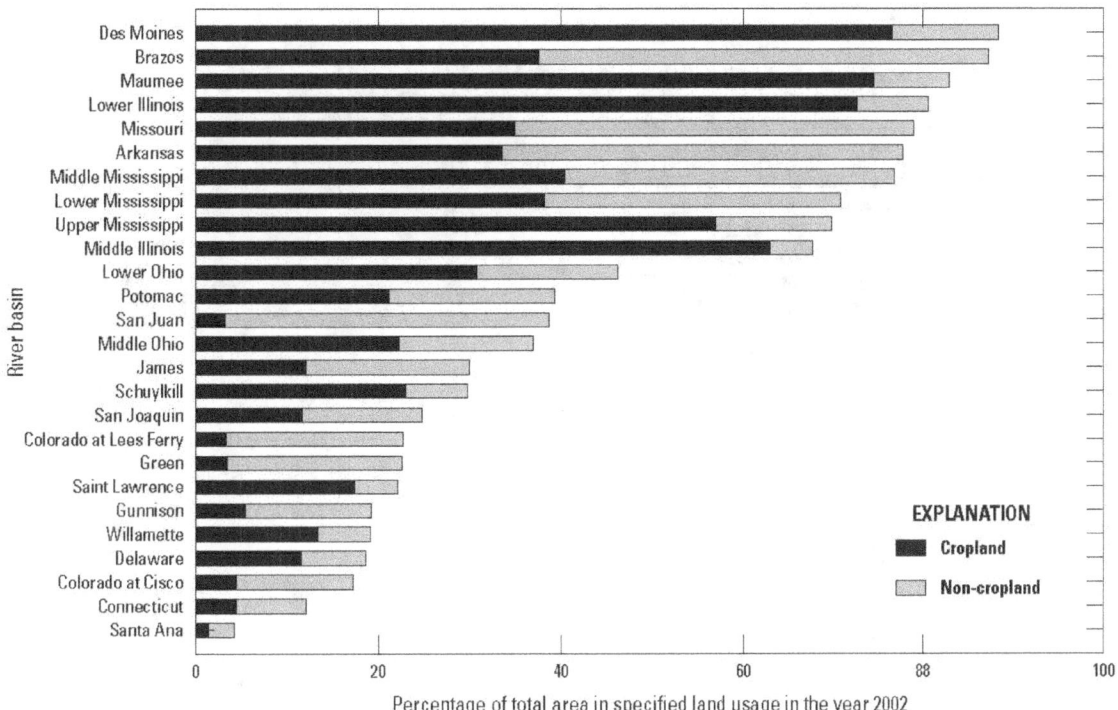

Figure 4.   Agricultural land use in 2002 for all river basins included in this study.

Cropland typically is more intensively managed, so basins where farmland is primarily cropland can be viewed as having a high degree of agricultural intensity. Ten of the river basins included in this study had greater than 60 percent of the basin area in farmland in 2002, four of the river basins exceeded 80 percent (fig. 4). There was a notable distinction among the most highly agricultural drainage basins based on whether or not the primary farmland usage was cropland. Western and Great Plains river basins, including the Brazos, Missouri, and Arkansas Rivers, had high percentages of total farmland and in each case, more than one-half of the farmland was non-cropland (fig. 4). However, several river basins located in the Midwestern United States had high percentages of the basin in farmland and almost all of this was cropland. These include the Des Moines, Maumee, Middle Illinois, Lower Illinois, and Upper Mississippi River basins (fig. 4). The Lower Mississippi River integrates inputs from many of the most heavily agricultural areas and had more than 60 percent of the basin in farmland, with the primary uses split between cropland and non-cropland (fig. 4). The remaining basins were all less than 50 percent farmland in 2002. However, even among these basins, a similar pattern remains with rivers in the Western United States, particularly those in the Colorado River basin with non-cropland as the primary agricultural land use while the rivers in the Eastern United States have primarily cropland usage (fig. 4).

Non-Point Sources of Nitrogen and Phosphorus from Fertilizer and Livestock

Nonpoint sources of nitrogen (N) from fertilizer and animal livestock were highest in the Des Moines River basin averaging more than 8 grams of nitrogen per square meter per year (g (N/m²)/yr). These sources also contributed more than 5 g (N/m²)/yr in the Maumee, Middle Illinois, Lower Illinois, and Upper Mississippi (fig. 5A). In contrast, nonpoint sources from fertilizer and livestock contributed less than 1 g (N/m²)/yr in the Connecticut, Delaware, Saint Lawrence, and all selected river reaches in the Colorado River basin over the same time period (fig. 5A). Fertilizer sources of N across all selected river basins ranged from less than 0.1 to almost 7 g (N/m²)/yr whereas animal sources ranged from about 0.2 to 2 g (N/m²)/yr. Fertilizer sources of N were about three times larger than animal sources of nitrogen in the river reaches with the highest inputs although animal sources dominated in the river reaches with the lowest sources of N (fig. 5A).

Phosphorus (P) sources followed largely the same geographic distribution with the Des Moines, Maumee, Upper Mississippi, Middle Illinois, and Lower Illinois all having greater than 1.0 g (P/m²)/yr inputs from animal livestock and fertilizer (fig. 5B). The river reaches with the lowest sources were the Saint Lawrence, Connecticut, and all river

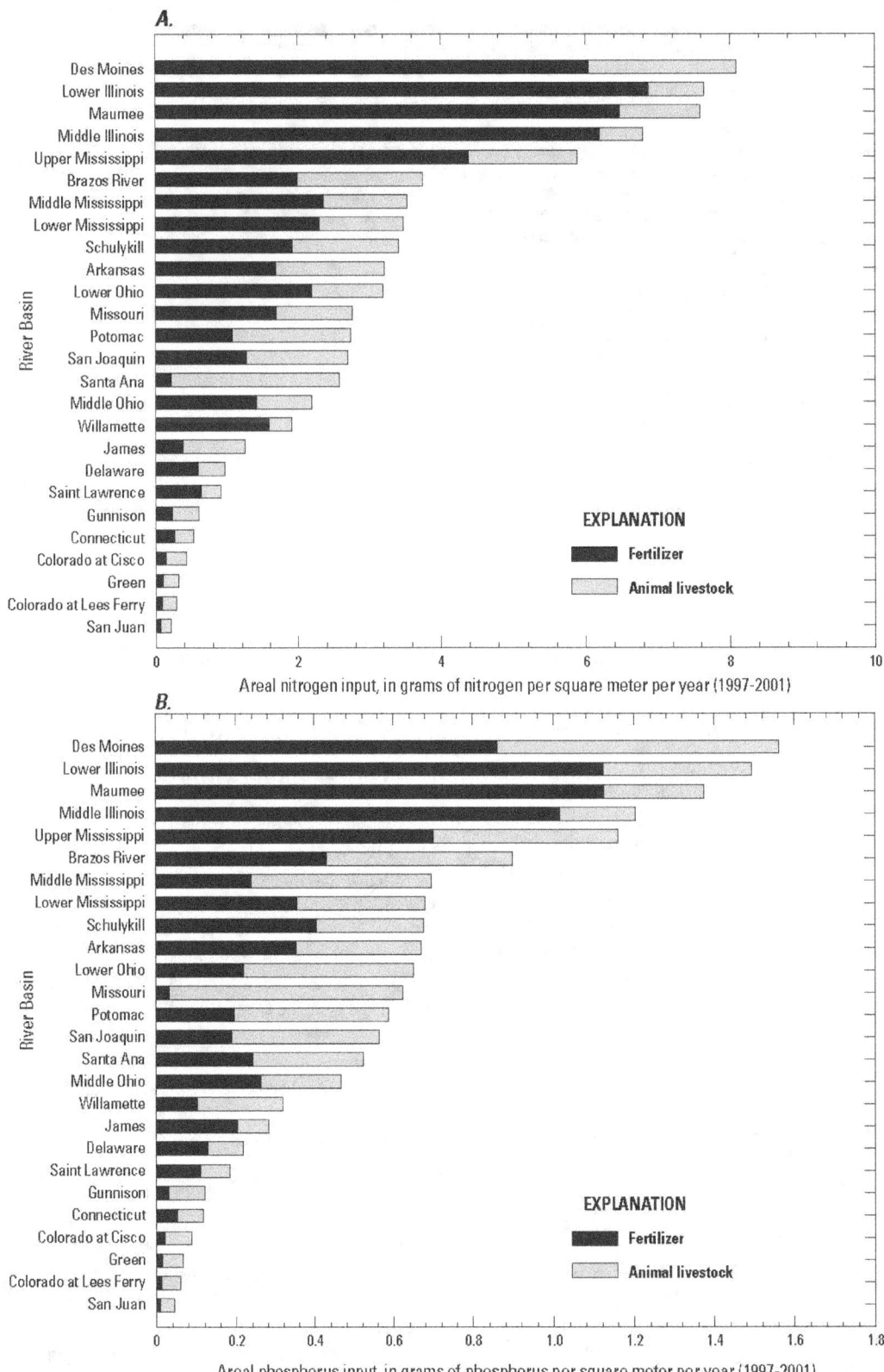

Figure 5.   Average nonpoint sources of (*A*) nitrogen and (*B*) phosphorus from fertilizer and animal livestock for all river basins included in this study, 1997–2001.

reaches of interest in the Colorado River basin, with less than 0.2 g (P/m²)/yr (fig. 5B). Animal livestock accounted for a much larger share of the nonpoint phosphorus that was identified in these basins and exceeded fertilizer sources of phosphorus in 14 of the basins (fig. 5B). Overall, animal livestock and fertilizer contributions had similar ranges, from approximately 0 to 1.0 g (P/m²)/yr (fig. 5B).

The highest fertilizer application rates were in the basins with the greatest cropland coverage, which emphasized the importance of chemical fertilizers to agricultural production in these areas. The difference in agricultural production between Great Plains river basins, exemplified by the Arkansas River, and the central and Mid-Western streams, the Illinois and Maumee River basins, are especially apparent.

Corn Production

Corn production in the river basins of interest was also analyzed as a more direct indicator of agricultural intensity. Averaged over 2000–2010, corn harvest ranged from less than 0.1 to more than 16,000 bushels/km² of drainage area (fig. 6). The intensively cultivated river basins in the Midwestern United States had the highest corn production, the Illinois, Des Moines, Maumee, and main stem Mississippi River basins (fig. 6). River basins with the lowest corn production were mostly in the West including all five drainage basins of interest within the Colorado River basin, the Willamette, Santa Ana, and San Joaquin Rivers (fig. 6). The James River basin in Eastern United States also had notably low corn production (fig. 6).

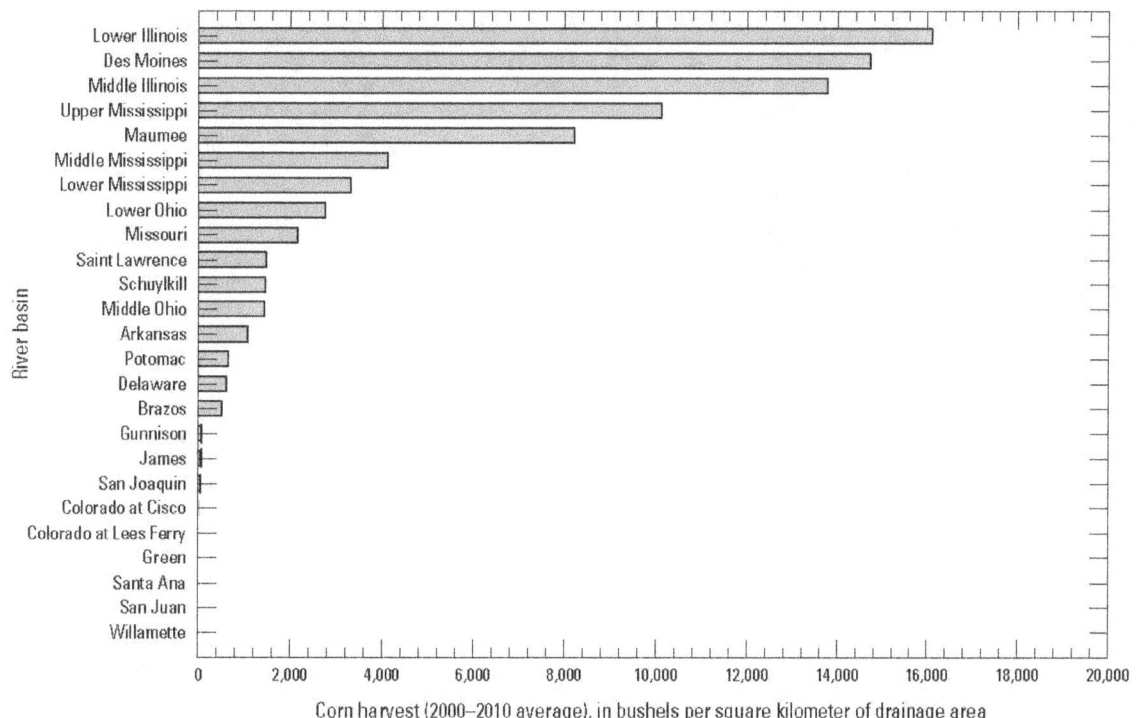

Figure 6. Average corn harvested per square kilometer of drainage area for 2000–2010 for all river basins included in this study.

Specific Basins of Interest

The following discussion is organized geographically, with basins grouped together roughly by the USGS system of hydrologic units (Seaber and others, 1987).

Eastern Basins

The eastern basins include streams in New England and the Mid-Atlantic physiographic provinces: the Connecticut River, the Delaware River, Schuylkill River (major tributary of the Delaware River), the Potomac River, and the James River. The Schuylkill River basin, which includes parts of Philadelphia and its western suburbs, was one of the most densely populated basins in our study, with more than 300 people/km² (fig. 3). The Schuylkill River basin stands out due to the high population densities as far back as 1890. Because the City of Philadelphia has long been a population center, population density in this basin had surpassed 100 people/km² by 1900 and has been increasing steadily since then (fig. 7A). In contrast, population density is less than the national average in the James River basin, although it has been increasing in the past several decades, doubling from 14 to 28 people/km² during 1980–2000. This increase reflects the relatively high population growth rates that were observed in the Mid-Atlantic during that time period.

The eastern basins generally have low areal percentages in farmland and cropland usage at present (figs. 4 and 7C–7D). These basins have steadily lost agricultural land throughout the 20th century (figs. 7C–7D). Most of these river basins were at least 70 percent agricultural land in 1900, but by 2002, none of exceeded 40 percent; the Connecticut and Delaware River basins were each less than 20 percent farmland (fig. 7C). This pattern reflects large-scale trends in agriculture throughout the 20th century, which was marked by specialization and consolidation resulting in a concentration of agricultural production in highly productive areas in the Central United States (Broussard and Turner, 2009). The loss of agricultural land in eastern basins also was reflected in the relatively low levels of fertilizer usage (fig. 5) and corn production, especially in the James River basin (fig. 6). Large increases in corn yields have caused corn production to increase in the Schuylkill River basin, despite the precipitous decrease in agricultural land in this basin (fig. 7D). However, the Schuylkill and Potomac Rivers also had fairly large inputs of nitrogen and phosphorus from animal livestock (figs. 5A–5B), which sets them apart from other basins in this region in terms of the amount of nitrogen and phosphorus from identified nonpoint sources (figs. 7E–7F).

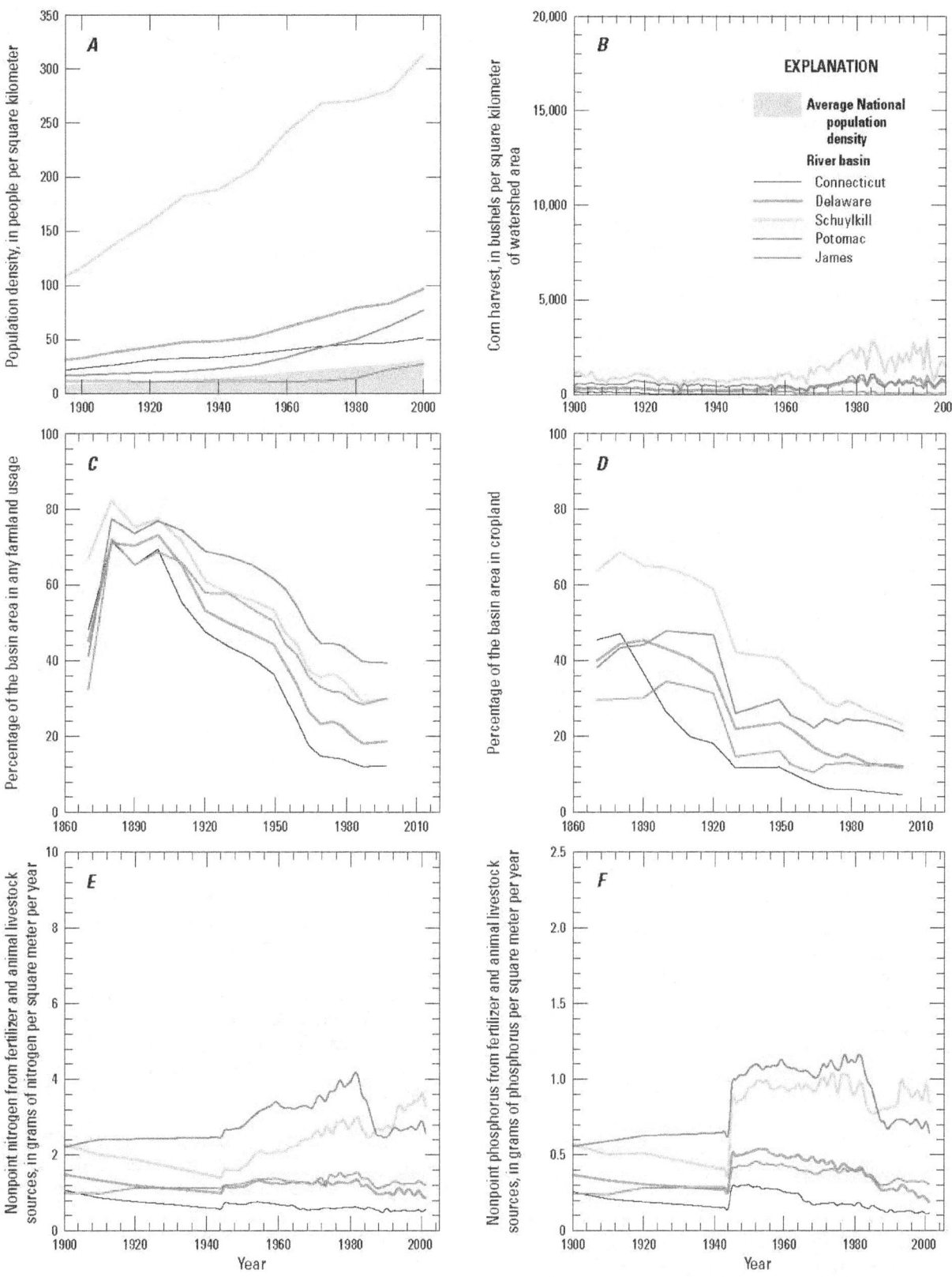

Figure 7.   Historical changes in (*A*) population density, (*B*) corn harvest, (*C*) percentage of basin area in any farmland usage, (*D*) percentage of basin area in cropland, and fertilizer and animal livestock sources of (*E*) nitrogen and (*F*) phosphorus in the Eastern basins.

Connecticut River

The Connecticut River is the longest river in New England, originating in the Canadian Province of Quebec, and northern New Hampshire and flowing south along the border of Vermont, through central Massachusetts and coastal Connecticut into Long Island Sound (fig. 8). The river channel descends rapidly in the upper reaches, becoming relatively flat and meandering for the remainder of its course (Benke and Cushing, 2005). The mouth of the river includes a long estuary reach, with tidal influence extending approximately 90 km (60 mi) upstream to Windsor Locks, Connecticut, located approximately 5 miles downstream of the Thompsonville monitoring station (fig. 8). The drainage area is about 25,000 km² (9,660 mi²) at the USGS gage at Thompsonville (station No. 01184000), just 6 mi upstream of the head of tide. At this point, mean annual discharge (1928–2008) is just more than 485 m³/s (17,000 ft³/s) and basin runoff averages more than 600 mm/yr. The basin is entirely within the New England physiographic province (fig. 2) and is largely forested, encompassing the New England Forests ecoregion in the north and the Northeastern Coastal Forests ecoregion in the south (Benke and Cushing, 2005). Climate generally is humid and temperate in character, with annual precipitation ranging from 165 cm (65 in.) at high elevations to 86 cm (34 in.) in the lowlands (Garabedian and others, 1998). Streamflow in the Connecticut River shows a strong seasonal pattern characteristic of snowmelt, with highest flows during spring and critically low flow during the summer months (Garabedian and others, 1998). Flow is heavily regulated by at least 125 reservoirs used for power generation and an additional 16 reservoirs focused on flood control (Garabedian and others, 1998).

The Connecticut River has likely been populated by humans since the retreat of the glaciers about 9,000 years ago (Benke and Cushing, 2005). Settlement by Europeans began when the Plymouth Colony was established in the lower river in 1633, with the river providing a key transportation corridor to the interior of the basin (Tetra Tech, 2000). Over the course of the next 200 years, about 75 percent of the forest in the basin was logged and converted to small-scale agricultural use, although most of the forest cover has been regained as agricultural usage has decreased (Tetra Tech, 2000). The northern part of the basin is currently mostly rural and relative wilderness with limited agriculture, although more urbanized areas occur farther downstream, especially focused in the tidal reach of the lower river. Some of these urban areas are among the oldest industrial areas in the Nation (Trench, 2000).

Water-quality issues in the Connecticut River have been of concern throughout the 20th century, initially focused on decreasing oxygen conditions in the lower river (Tetra Tech, 2000). As early as 1897, the Connecticut General Assembly recognized the problem of pollution in the river and initiated an investigation of sewage-disposal practices (Hupfer, 1965). Despite this attention to the issue, however, by the 1930s, only about one-third of wastewater effluent received some form of primary treatment (Mullaney, 2004). In 1955, the New England Interstate Water Pollution Control Commission determined the reach of the river from Holyoke Dam in Massachusetts to Hartford, Connecticut (about 20 mi downstream of Thompsonville) was a Class D waterway, only suitable for transportation of sewage and industrial waste (Tetra Tech, 2000). The severe water pollution observed in this reach was due to effluent from industry and municipal sewage, much of which was not treated prior to the Clean Water Act of 1972 (Tetra Tech, 2000). The primary industries responsible for water-quality degradation included paper mills, chemical, metal, plating, and dyeing (Mullaney, 2004). With the implementation of advanced wastewater treatment during the 1970s and 1980s, dissolved oxygen conditions significantly improved, although combined sewer overflows continue to discharge untreated sewage to the river during storms (Tetra Tech, 2000).

The population density in the basin upstream of the tidal reach was estimated to be about 52 people/km² in the year 2000 (fig. 3; U.S. Census Bureau, 2000). This value reflects the influence of the largely unpopulated region in the northern part of the basin. Nitrogen and phosphorus inputs from fertilizer and animal livestock averaged 0.5 and 0.1 (g/m²)/yr, respectively, and were split almost evenly between the two sources (fig. 5). Major nutrient sources to the Connecticut River also include municipal wastewater discharges and nonpoint runoff from urban areas and agricultural activities (Trench, 2000). Additionally, the widespread urban development across the Northeastern United States results in significant deposition of nitrogen to the landscape from fossil fuel deposition (Jaworski and others, 1997).

Data Sources

Data sources for the Connecticut River are summarized in table 4. Mean daily streamflow data are available for the reference stream gaging station at the Connecticut River at Thompsonville, Conn. (station No. 01184000) since 1928. USGS data for nitrogen are available for the Connecticut River at Warehouse Point, Conn., for 1890–1899; and for the Connecticut River below Springfield, Mass., from 1900. Nitrogen data from the Massachusetts State Department of Health also are available for the same site for 1914–1918. This collection site is downstream of the sewer outfall located near Westfield River in Springfield, Mass., and reflects the influence of this urban area. Works Progress Administration water-quality data are available for the Connecticut River below Thompsonville, Conn., for 1937. USGS data for the Connecticut River at Thompsonville, Conn., are available beginning in 1952.

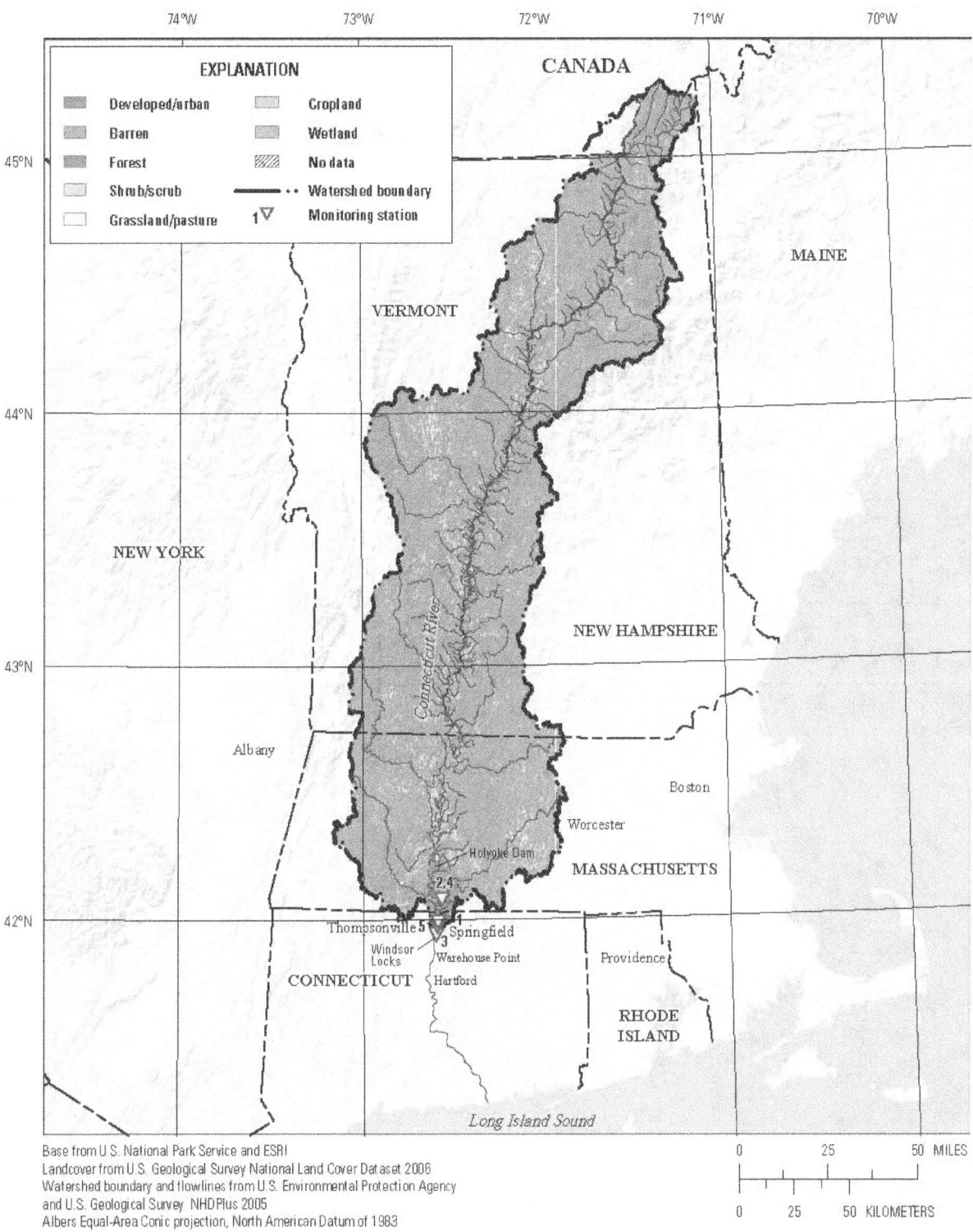

Figure 8.   Connecticut River basin.

Table 4. Connecticut River basin data sources.

[**Monitoring station** is shown in figure 8. **Source:** NWIS, National Water Information System; WPA, Works Progress Administration. **Constituent:** Alk, alkalinity. See table 2 for definitions of all other constituent abbreviations]

Monitoring station	Source	Station	Latitude	Longitude	Constituent	Start year	End year
[1]1	NWIS	01184000 Connecticut River at Thompsonville, Conn.	41°59'14"	72°36'19"	Daily streamflow	1928	2008
					TN	1969	2009
					NH$_3$	1969	2009
					TON + NH$_3$	1969	2009
					NO$_3$	1952	2009
					Alk	1952	2009
					TP	1968	2009
					TDP	1966	2009
[2]2	Leighton, 1903	Connecticut River below Springfield, Mass.	42°04'54"	72°34'58"	NH$_3$	1900	1900
					NO$_3$	1900	1900
[2]3	Leighton, 1903	Connecticut River at Warehouse Point, Conn.	41°56'37"	72°36'50"	NH$_3$	1890	1899
					NO$_3$	1890	1899
[2]4	Massachusetts State Department of Health, Fourth Annual Report	Connecticut River below Springfield, Mass.	42°04'54"	72°34'58"	NH$_3$	1914	1918
					NO$_3$	1914	1914
[2]5	WPA, 1940	Connecticut River below Thompsonville, Conn.	41°58'30"	72°36'23"	NH$_3$	1937	1937
					NO$_3$	1937	1937
					Alk	1937	1937

[1]Reference stream gaging station.

[2]Historical station, location approximate.

Delaware River

The Delaware River basin has a long history of European settlement, dating back to the colonial period of the 17th century (Patrick, 1992). The river originates in the Catskill Mountains and flows from north to south, draining parts of New York, Pennsylvania, New Jersey, and Delaware, before emptying into the Atlantic Ocean (fig. 9). The length of river of interest for this analysis is the non-tidal reach above the Delaware Estuary that heads at Trenton, New Jersey. The target drainage area encompasses 17,600 km^2 (about 6,800 mi^2), excluding the Schuylkill River basin that drains the city of Philadelphia farther downstream and is considered separately in this report. The river flows through complex terrain, incorporating five physiographic provinces: Appalachian Plateau, Valley and Ridge, New England, Piedmont Plateau, and Coastal Plain (fig. 2). The Lehigh River is a major tributary that drains important coal and steel producing areas of the Great Appalachian Valley. Climate is humid and continental, with little variation in precipitation across the drainage area (averaging about 106 cm/yr or 42 in/yr), although extreme precipitation occurs infrequently as hurricanes (Benke and Cushing, 2005). Mean annual discharge (1912–2008) is approximately 330 m^3/s (11,800 ft^3/s), so that average runoff is about 600 mm/yr.

When first discovered by Henry Hudson in 1609, the Delaware River basin was covered with forests, and timber harvest quickly became a major industry supplying wood for paper and building ships (Patrick, 1992). Abundant fishes in the river and the estuary supported important commercial fisheries during the 17th and 18th centuries, and the Delaware Valley grew to become a key center of trade. Cities were established along the lower river, especially after municipal water systems were developed to utilize river water for domestic and industrial use. Factories proliferated to utilize the availability of coal and iron, and glassblowing, silkweaving, pottery, and porcelain, as well as iron shipbuilding and munitions, became important industries during the 19th century (Patrick, 1992). The growth of factories in the lower Delaware Valley was associated with further increases in population during this time in the urban areas, especially in Philadelphia. Agriculture expanded throughout the middle reaches of the basin to meet the increasing demand for food. Numerous canals were built to provide a transportation network to support the growing population (Benke and Cushing, 2005). By the beginning of the 20th century, important water-quality issues were related to untreated sewage and industrial waste discharged from cities that simultaneously drew water from the river for their water supply.

Attempts at controlling pollution by sewage treatment during the first half of the 20th century were largely offset by ongoing growth of cities and industries (Patrick, 1992). The large quantities of waste material created significant oxygen demand, threatened industrial production, and contributed to abrupt decrease in commercial fishing (Patrick, 1992). Water quality in the lower river was further impacted by impoundment of headwater streams in the 1920s to augment the water supply for New York City, although water levels are now carefully regulated to maintain flow requirements to minimize detrimental downstream impacts (Fischer and others, 2004). In 1936, the Interstate Commission on the Delaware River Basin (INCODEL) was established to improve water-quality conditions (Kiry, 1974). INCODEL was incorporated with the newly formed Delaware River Basin Commission (DRBC) in 1955, which focused on a wide range of environmental problems in the basin, including flood control, water quality, and water supply (Kiry, 1974). As a result of these efforts during the 1950s, sewage treatment facilities began to be built and minimal treatment of industrial wastes became more common. By the end of that decade, the proportion of cities with "adequate" sewage treatment plants increased from 20 to 75 percent, most conducting secondary treatment (Albert, 1982). Substantial improvements in water quality ensued, especially as industrial loads were further reduced in response to passage of the Clean Water Act in 1972.

Contamination from fracking procedures associated with natural gas drilling is an important recent concern in the Delaware River basin. The Marcellus Shale is the largest shale basin in the continental United States (Kargbo and others, 2010), and underlies about 36 percent of the Delaware River basin (Delaware River Basin Commission, 2011). The Delaware River Basin Commission has been in the forefront of the development of regulations to limit the environmental impacts of fracking on water quality, especially because the Delaware River serves as a critical source for drinking water.

Presently, there exists a strong gradient from north to south in land use and population pressure in the Delaware River basin. Although, the northern part of the basin remains largely forested, only a small percentage of the basin population lives there (Fischer and others, 2004). Upstream of the Delaware Water Gap, two reaches of the river have been designated as Scenic and Recreational Rivers, which reflects the high quality of water and stringent protection provided (Albert, 1982). Farther downstream, agriculture remains an important influence in the basin, with corn and soybeans the most widely grown crops (Hickman, 2004). However, basinwide fertilizer and animal livestock sources of nitrogen and phosphorus are very low. Nitrogen inputs averaged less than 1.0 g (N/m^2)/yr from 1997 to 2001 and phosphorus

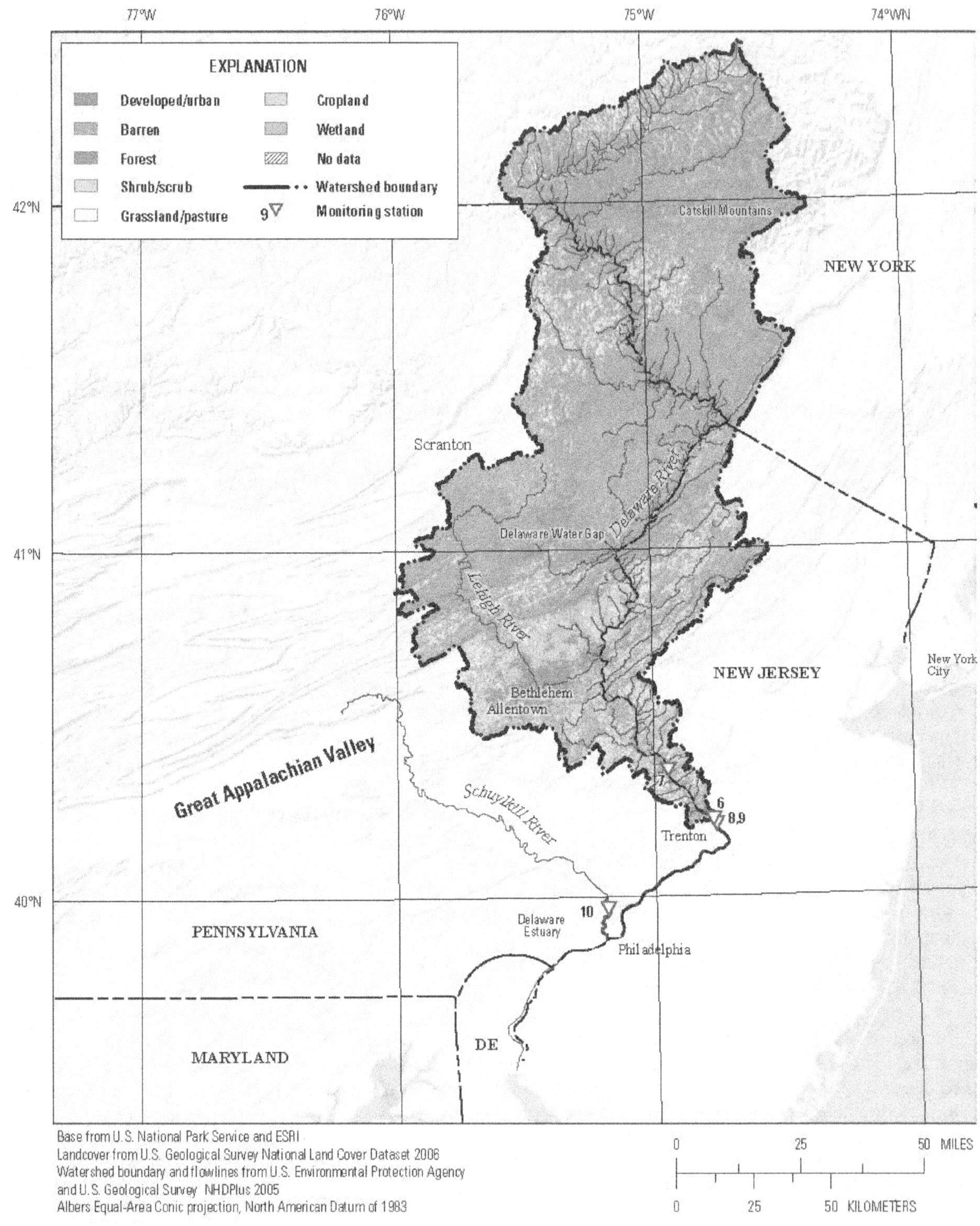

Figure 9.   Delaware River basin.

inputs were 0.2 g (P/m^2)/yr in the same time period. Fertilizer sources of nitrogen and phosphorus were approximately twice as large as animal livestock sources (fig. 5). The most significant water-quality impact on the lower Delaware River basin is from urbanization and industrial activity, including the influence of the lower Lehigh River that drains the highly industrialized eastern Pennsylvania (Albert, 1982). Nonetheless, despite the significant urbanization in the lower basin, the population density for the entire basin upstream of Trenton is only moderately high compared to other basins in this study, estimated as 97 people/km^2 in the year 2000 (fig. 3; U.S. Census Bureau, 2000).

Data Sources

Available data sources for the Delaware River are summarized in table 5. Daily streamflow data are available from the reference stream gaging station at Trenton since 1912 (station No. 01463500). Water-quality data are available from the Delaware River at Trenton for 1900, 1923, and from 1944 through 2009. Additional water-quality data are available from the Delaware River at Lambertville from 1906 and 1976–82; the overlap in time for data from this site and from Trenton allows an evaluation of comparability.

Table 5.   Delaware River data sources.

[**Monitoring station** is shown in figure 9. **Source:** NWIS, National Water Information System. **Constituent:** Alk, alkalinity. See table 2 for definitions of all other constituent abbreviations]

Monitoring station	Source	Station	Latitude	Longitude	Constituent	Start year	End year
[1]6	NWIS	01463500 Delaware River at Trenton, N.J.	40°13'18"	74°46'41"	Daily streamflow	1912	2008
					TN	1969	2009
					NH$_3$	1970	2009
					TON + NH$_3$	1969	2009
					NO$_3$	1944	2009
					Alk	1944	2009
					TP	1969	2009
					TDP	1964	2009
7	NWIS	01462000 Delaware River at Lambertville, N.J.	40°21'53"	74°56'56"	Daily streamflow	1897	1906
					TN	1976	1982
					NH$_3$	1976	1982
					TON + NH$_3$	1976	1982
					NO$_3$	1906	1982
					Alk	1906	1982
					TP	1976	1978
[2]8	Leighton, 1903	Delaware River at Trenton, N.J.	40°12'28"	74°46'03"	NH$_3$	1900	1900
					NO$_3$	1900	1900
					Alk	1900	1900
[2]9	Collins and Howard, 1928	Delaware River at Trenton, N.J.	40°12'28"	74°46'03"	NO$_3$	1923	1924
					Alk	1923	1924

[1]Reference stream gaging station.

[2]Historical station, location approximate.

Schuylkill River

The Schuylkill River is the largest tributary to the Delaware River, and located entirely within the State of Pennsylvania (fig. 10). The river begins in the Blue Mountain part of the Valley and Ridge physiographic province of the Appalachian Mountains (fig. 2). From there it flows through the Piedmont Plateau and Coastal Plain provinces before bisecting the city of Philadelphia and emptying into the Delaware River Estuary (Stamer and others, 1985). The drainage area for the non-tidal river upstream of the Estuary, referenced at the USGS gage in Philadelphia, is about 4,980 km² (1,900 mi²). Climate in the basin generally is humid and precipitation is distributed evenly throughout the year, ranging from approximately 114–127 cm/yr (45–50 in/yr) in the high elevation headwater regions to about 109 cm/yr (43 in/yr) at the mouth (Schuylkill Watershed Conservation Plan, 2001). Mean annual discharge (1931–2008) for the gage at Philadelphia during the 20th century is approximately 77 m³/s (2,700 ft³/s), which corresponds with mean annual runoff of nearly 500 mm/yr (19 in/yr).

The lower region of the Schuylkill River was first settled by Europeans when Philadelphia was founded in the early 17th century. Population growth was slow until William Penn received the charter from King Charles II to establish the Commonwealth of Pennsylvania in 1680, after which population density began to increase rapidly in the Philadelphia area (Schuylkill Watershed Conservation Plan, 2001). By 1750, Philadelphia was the largest city in the colonies, serving as an important center of trade and government. Early land use in the basin during the colonial period focused on farming, timber harvest, and development of the iron industry, leading to the growth of Philadelphia into the most populous city during the Revolutionary War. Agriculture in the basin was a key element of the colonial economy, especially focused on wheat, and southeastern Pennsylvania was recognized as the largest producer of food in the Nation until the middle of the 19th century (U.S. Department of the Interior, 1987).

During the 19th century, the Schuylkill River became a critical industrial corridor, facilitating the transportation of anthracite coal from the upriver Appalachian mines to the factories and mills established in the cities of the lower river reaches (Schuylkill Watershed Conservation Plan, 2001). The Schuylkill Navigation System was constructed, consisting of 32 dams and 103 locks, to facilitate the use of the river for transportation (Schuylkill Watershed Conservation Plan, 2001). At the same time, the river was dammed at the Fairmont Water Works in 1815 for use as municipal water supply by the city of Philadelphia and surrounding areas. Discharge of acidic wastes and fine sediment from coal mines caused clogging of the river channel and significant degradation of water quality, the effects of which peaked by the middle of the 20th century (Stamer and others, 1985). Water quality was further impacted by discharge of wastewater from the numerous cities located along the river. The decline in water quality was a subject of intense interest, resulting in passage of the Clean Streams Act and the Desilting Act in 1945 by the State of Pennsylvania, which led to removal of unprecedented volumes of sediment from the river (Stamer and others, 1985). By this time, agriculture in the basin was dominated by dairy farming and egg production, although the number of farms was declining due to expanding urbanization (U.S. Department of the Interior, 1987).

The cleanup of the Schuylkill River in the last half of the 20th century has been remarkable. In 1979, the lower Schuylkill River was the first river segment to be added to the Pennsylvania Scenic Rivers System, emphasizing the increase in recreational use of the river as well as protection of anadromous fish populations (Stamer and others, 1985). Nonetheless, the Schuylkill River remains heavily affected by human land use, especially from the high population density (314 people/km²) (fig. 3). Agriculture also is still a significant influence (Jaworski and others, 1997) (figs. 5A–5B), although the Schuylkill River basin has experienced decreases in farmland and cropland similar to other Eastern basins (figs. 7C–7D). Nitrogen and phosphorus sources from fertilizer and animal livestock are low compared to other basins included in this study (fig. 5), but are among the highest in the Eastern basins (figs. 7E–7F).

Data Sources

Data sources for the Schuylkill River are summarized in table 6. Daily streamflow data are available for the reference stream gaging station at the Schuylkill River at Philadelphia (station No. 0147500) beginning in 1931. Water-quality data are available for the same location beginning in 1925, depending on constituent, and extending through the 20th century.

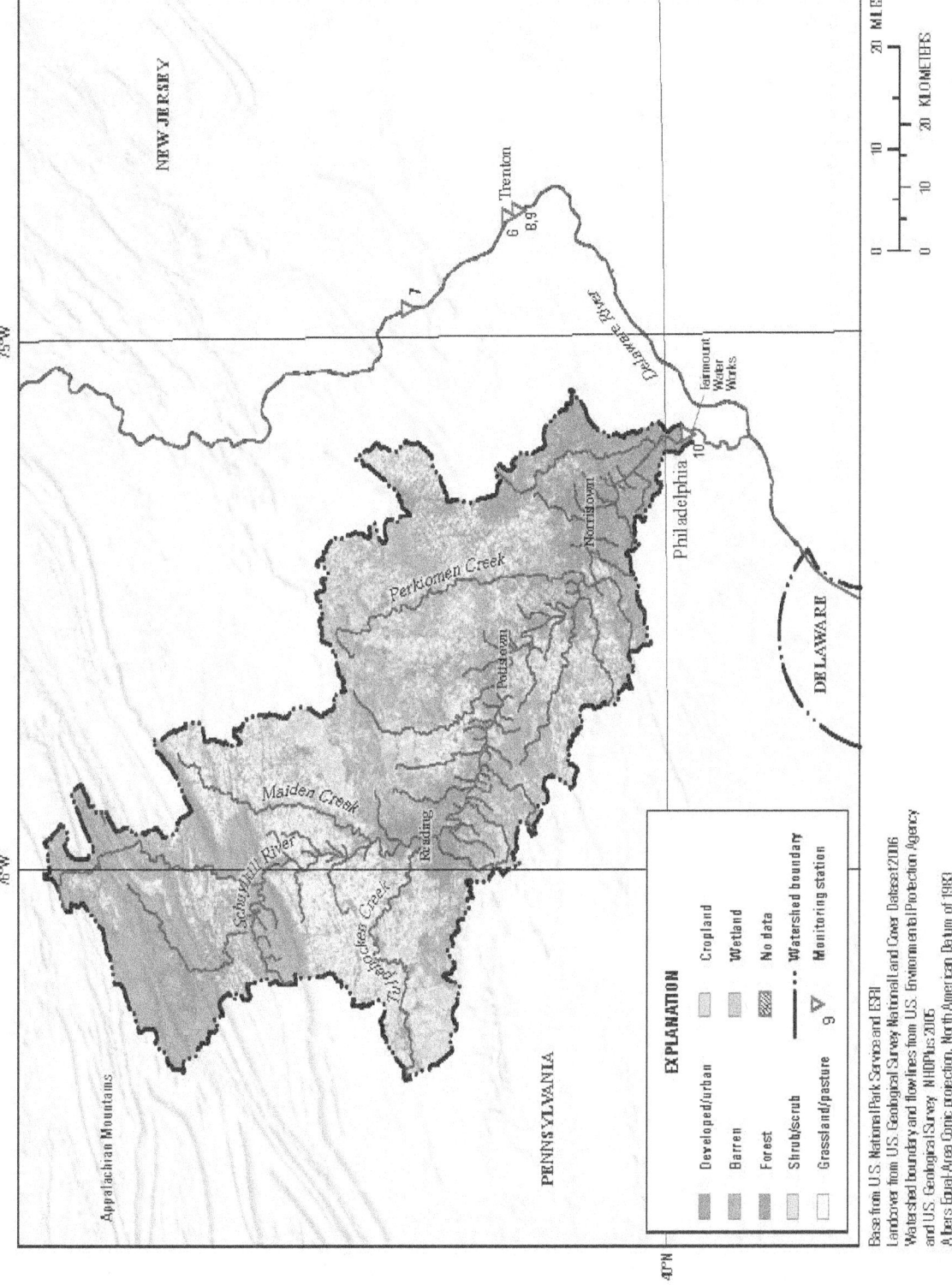

Base from U.S. National Park Service and ESRI
Landcover from U.S. Geological Survey National Land Cover Dataset 2006
Watershed boundary and flowlines from U.S. Environmental Protection Agency
and U.S. Geological Survey NHDPlus 2005
Albers Equal-Area Conic projection, North American Datum of 1983

EXPLANATION

Developed/urban		Cropland
Barren		Wetland
Forest		No data
Shrub/scrub	·—··	Watershed boundary
Grassland/pasture	▽ 9	Monitoring station

Figure 10.   Schuylkill River basin.

Table 6. Schuylkill River data sources.

[**Monitoring station** is shown in figure 10. **Source:** NWIS, National Water Information System. **Constituent**: Alk, alkalinity. See table 2 for definitions of all other constituent abbreviations]

Monitoring station	Source	Station	Latitude	Longitude	Constituent	Start year	End year
[1]10	NWIS	01474500 Schuylkill River at Phildelphia, Pa.	39°58'04"	75°11'20"	Daily streamflow	1931	2008
					TN	1954	2004
					NH_3	1970	2004
					$TON + NH_3$	1969	2004
					NO_3	1933	2004
					Alk	1925	2004
					TP	1969	2004
					TDP	1969	2004

[1]Reference stream gaging station.

Potomac River

The Potomac River is located on the mid-Atlantic seaboard and drains parts of the Appalachian Mountains, piedmont, and coastal plain in Virginia, West Virginia, Maryland, and Pennsylvania (fig. 11). It is the second largest river flowing into Chesapeake Bay with a drainage area of 37,800 km² (14,600 mi²). For the purposes of this study, data were considered from the reference stream gaging station located just upstream of Washington, D.C. (station No. 01646502). At this point, the drainage area is 29,900 km² and mean annual discharge is 337 m³/s (11,888 ft³/s). Annual average discharge ranges from approximately 140 to 680 m³/s (5,000 to 24,000 ft³/s) throughout the period 1930–2009. Runoff averages 355 mm/yr throughout the same period but also is highly variable ranging from 140 to 730 mm/yr. This region is susceptible to large cyclonic storms originating in the Atlantic Ocean, partially accounting for the large interannual variability in river flow.

The Potomac River originates in the Appalachian Plateau physiographic province and runs through portions of the Valley and Ridge, Piedmont Plateau, and Blue Ridge physiographic provinces (fig. 2) before entering the Coastal Plain near Washington, D.C. More than 70 percent of the total river discharge originates in the mountainous regions upstream of the Piedmont Plateau. The North Branch Potomac, South Branch Potomac, and Shenandoah Rivers are major upland tributaries to the Potomac River and the Monocacy River is a major tributary that joins the Potomac River in the Piedmont Plateau.

Land use in the Potomac River has changed dramatically throughout the 20th century. In 1900, 80 percent of the basin was agricultural land with approximately 50 percent of the basin in cropland (figs. 7E–7F). However, as with many areas in the Eastern United States, agriculture became much less prominent as large-scale farming consolidated within the Midwestern United States throughout the 20th century. At the last Census of Agriculture in 2002 (U.S. Department of Agriculture, 2005), less than 40 percent of the Potomac River basin was in farmland usage with 20 percent of the basin area used as cropland (fig. 4). Corn harvest was relatively low, 660 bushels/km² of drainage area (fig. 7B). Accordingly, fertilizer usage in the basin at the end of the 20th century is low compared with other basins of interest to this study. Nitrogen and phosphorus from fertilizer sources averaged only 1.1 and 0.2 (g/m²)/yr from 1997 to 2001, respectively (fig. 5). However, animal livestock sources of both N and P were larger than fertilizer sources, 1.7 and 0.5 (g/m²)/yr, respectively (fig. 5). Population density in the basin has increased during the 20th century from 17 people/km² in 1900 to more than 77 people/km² in 2000, reflecting the growth of metropolitan Washington, D.C. and suburbs (fig. 7A).

Coal mining activity has long been a concern for water quality in the Potomac River basin. The economic viability of resource extraction was greatly aided by the construction of the Chesapeake and Ohio Canal in 1850, which ran from Washington, D.C. to Cumberland, Md. Coal mining began with small underground mines in the early 19th century, which gradually gave way to larger consolidated underground mines (Jaworski and others, 2007). Surface coal mining is the predominant type of coal mine operating in the basin today, but these types of mines did not begin until 1920 and increased significantly after World War II (Mills and Davis, 2000).

By the late 19th century, concerns about pollution on the Potomac River were already being raised. The combination of population growth and the presence of heavy industry led to serious public health concerns. The Potomac River was one of the first basins studied by the U.S. Public Health Service in early 20th century because of its interstate nature, the predominance of pollution, and the fact that it served as the drinking water supply to Washington, D.C. (Cumming and others, 1916). In particular, there was a concern that the lethality of mining and industrial wastes inhibited the normal purification process in the river thus allowing pollution to persist far downstream from its source (Cumming and others, 1916). Intensely degraded conditions were described in some of the headwater streams where mining was prevalent, such as Georges Creek in western Maryland, where the waters were said to be "lethal instead of life-giving in character. They destroy all vegetable and animal life in the channel, and stain the rocks on which they flow yellow..." (Parker and others, 1907).

Environmental conditions in the Potomac River continued to deteriorate throughout most of the 20th century, eventually having a significant negative impact on the Chesapeake Bay. By mid-century, low oxygen in the river led to fish kills; silt pollution was considered a primary threat to the Chesapeake Bay; and much of the riverine and estuarine habitat was threatened (U.S. Department of the Interior, 1968; Federal Water Pollution Control Administration, 1969; Maryland Department of Natural Resources, 2007).

State and multi-State efforts to remediate the Potomac River began in the 1930s and 1940s. Actions by individual States included limiting acid mine drainage pollution in Pennsylvania in 1945 and the creation of the Water Pollution Control Commission by Maryland in 1947. In 1940, the U.S. Congress also established the Interstate Commission on the

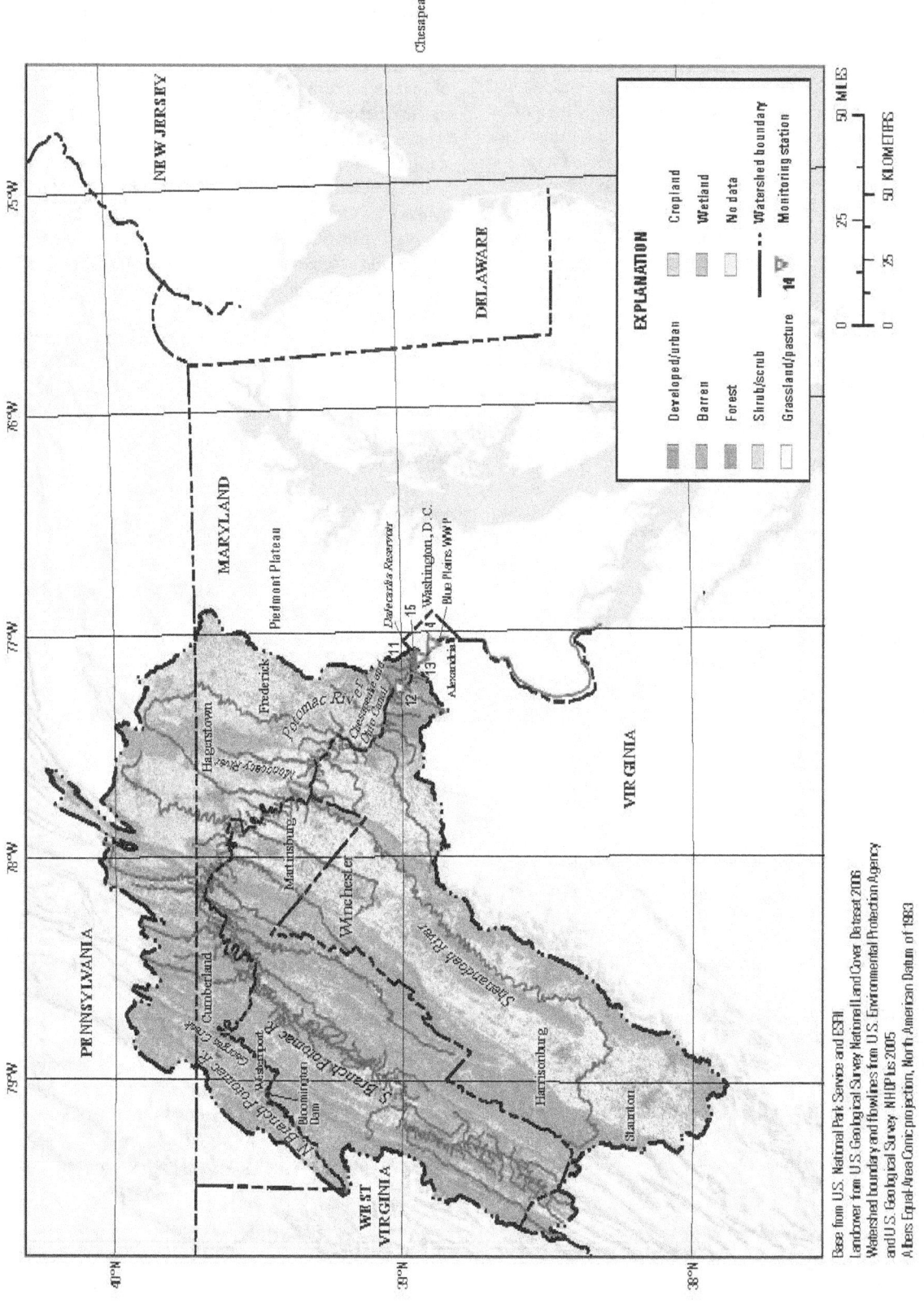

Base from U.S. National Park Service and ESRI
Landcover from U.S. Geological Survey National Land Cover Dataset 2006
Watershed boundary and flowlines from U.S. Environmental Protection Agency
and U.S. Geological Survey NHDPlus 2005
Albers Equal-Area Conic projection, North American Datum of 1983

Figure 11. Potomac River basin.

Potomac River Basin (ICPRB) to address pollution in the Potomac River. The commission consisted of members from Pennsylvania, Maryland, West Virginia, Virginia, and the District of Columbia. Other interstate efforts included the Potomac Enforcement Conference, which convened in 1957 and the Potomac River Basin Advisory Committee, which was established in 1965. The first wastewater-treatment plant was built in 1938 downstream of Washington, D.C. to focus on reducing pollution in the lower river and the estuary. Other treatment plants were built in the 1950s and 1960s at Alexandria, Va., Cumberland, Md., and Westernport, Md. Improvements also were made to connect three suburban counties (Fairfax, Loudon, and Montgomery) to the Blue Plains wastewater-treatment plant. Thus, by 1968, about 85 percent of all municipal and industrial effluent in the basin received "some degree of treatment" (U.S. Department of the Interior, 1968). By 1980, all wastewater-treatment plants in the basin had secondary treatment in operation (Jaworski et al. 2007).

Actions also have been taken to reduce acid mine drainage pollution in the basin. Bloomington Dam was constructed by U.S. Army Corps of Engineers on the North Branch Potomac River by 1981 partly with the goal of reducing downstream impacts of acid mine drainage. In 1993, Maryland and West Virginia entered an agreement with the ICPRB to restore water quality in the North Branch Potomac River by neutralizing acidic waters from mine drainage (Mills and Davis, 2000). Some estimates show that acid loading to the Potomac River had decreased by almost 90 percent between 1940 and 1988 (Mills and Davis, 2000).

Environmental conditions in the Chesapeake Bay came under scrutiny in the 1970s as pollution caused noticeable degradation of its fisheries and ecological health. In 1983, the Chesapeake Bay Program was formed to address pollution problems pertaining to the Chesapeake Bay. The Potomac River was among the basins targeted for pollution reductions. Principal objectives of the program included reductions in nitrogen and phosphorus pollution to the bay (Sprague and others, 2000). By 1998, the USGS estimated that landscape best-management practices had reduced the agricultural load of nitrogen to the Potomac River by 9 percent since 1985 (Sprague and others, 2000).

Data Sources

Data sources for the Potomac River are summarized in table 7. Daily streamflow data are available beginning in 1930 at the reference stream gaging station at the Potomac River near Washington, D.C. (station No. 01646502. As shown in table 7, the earliest known reliable data on nitrogen in Potomac River were collected by the USGS at this site in 1921. The U.S. Army Corps of Engineers also analyzed the water drawn from the Potomac River at Great Falls at Dalecarlia Reservoir water intake (monitoring station 15), which supplies Washington, D.C. Records of these analyses are available from 1922 through the mid-1960s (with some gaps, particularly in the 1940s). The USGS has been analyzing Potomac River waters at various intervals since 1973.

Table 7.   Potomac River data sources.

[**Monitoring station** is shown in figure 11. **Source:** NWIS, National Water Information System; STORET, Storage and Retrieval Data Warehouse. **Constituent:** Alk, alkalinity. See table 2 for definitions of all other constituent abbreviations]

Monitoring station	Source	Station	Latitude	Longitude	Constituent	Start year	End year
[1]11	NWIS; Clarke, 1924	01646502 Potomac River (adjusted) near Washington, D.C.	38°56′58″	77°07′40″	Daily streamflow NO$_3$ Alk	1930 1921 1921	2009 1921 1921
12	NWIS	01645500 Potomac River at Great Falls, Md.	39°00′03″	77°14′55″	TN NH$_3$ TON + NH$_3$ NO$_3$ Alk TP TDP	1980 1973 1980 1980 1973 1980 1980	1980 1980 1980 1980 1973 1980 1980
13	NWIS	01646580 Potomac River at Chain Bridge, at Washington, D.C.	38°55′46″	77°07′01″	TN NH$_3$ TON + NH$_3$ NO$_3$ Alk TP TDP	1973 1973 1973 1973 1973 1973 1973	2009 2009 2009 2009 2009 2009 2009
14	STORET	100130 Potomac River at Washington, D.C.	38°53′14″	77°03′20″	TN NH$_3$ TON + NH$_3$ NO$_3$ Alk TP TDP	1964 1963 1964 1964 1963 1964 1964	1969 1965 1969 1969 1968 1969 1969
15	District of Columbia water system	Potomac River at Dalecarlia water intake, near Washington, D.C.	38°55′57″	77°07′02″	NH$_3$ NO$_3$ Alk	1922 1922 1922	1928 1965 1965

[1]Reference stream gaging station.

James River

The influence of European settlement in the James River basin began with establishment of Jamestown in the James River Estuary in 1607 (Benke and Cushing, 2005). It is the largest river contained entirely within the Commonwealth of Virginia, with the headwaters in the mountains along the western border and flowing through central Virginia to empty into the Chesapeake Bay (fig. 12). The target drainage area for this study is defined by the USGS reference stream gaging station at Cartersville, measured as approximately 16,200 km² (6,250 mi²), located about 40 mi upstream of the head of tide in Richmond (Belval and others, 1994). The river crosses three physiographic provinces as it flow from west to east: the Valley and Ridge, Blue Ridge, and Piedmont Plateau physiographic provinces. Climate is relatively mild, with hot, humid summers, and cool winters. Severe weather is a concern, including occasional heavy rain and high winds from hurricanes. Precipitation averages about 108 cm/yr (42 in/yr) (Benke and Cushing, 2005). Mean annual discharge at Cartersville for the period 1899–2008 was approximately 200 m³/s (7,100 ft³/s), which results in mean annual runoff of about 390 mm/yr.

At the turn of the century, the dominant mode of agricultural activity in the Appalachian and Piedmont regions of the James River basin was subsistence farming (Pudup, 1990). An important component of the human economy in this region during this time also was provided by the abundant chestnut forests in the Blue Ridge region of the upper basin. These areas served as an important source for timber as well as the trade in chestnuts, which were transported via the railroad to urban centers, such as Philadelphia and New York (Lutts, 2004). These activities declined rapidly by the 1930s coincident with the obliteration of the chestnut forests due to the invasion of the Chestnut fungus.

Although the James River drainage basin contains large municipal areas in the lower reaches, the portion of the basin targeted by this study remains largely rural with only a few cities with population greater than 50,000 people. Population density is estimated at about 27 people/km², based on the 2000 census (fig. 3; U.S. Census Bureau, 2000). Most of the basin is forested, although agriculture is still recognized as an important influence on water quality (Benke and Cushing, 2005). Over the course of the 20th century, farmland acreage and the total number of farms has decreased significantly in a pattern similar to that observed throughout the Eastern United States (Cohen, 2009) (figs. 7C–7D). Currently, agricultural lands comprise 15 percent of the basin upstream of Cartersville (Sprague and others, 2009) (fig. 4). Animal sources of nitrogen and phosphorus were more than double that of fertilizer on average from 1997 to 2001 (fig. 5). However, the sum of these sources of only 1.3 g (N/m²)/yr and 0.3 g (P/m²)/yr (fig. 5), and has been steady or decreasing for most of the late 20th century (figs. 7E–7F). Since 1983, with the establishment of the Chesapeake Bay Program, there has been a strong emphasis on improving agricultural practices throughout the James River basin to reduce loads of nutrients and sediment to Chesapeake Bay (Virginia Department of Environmental Quality, 2005).

Data Sources

Data sources for the James River are summarized in table 8. Daily streamflow data are available from the reference stream gaging station, James River at Cartersville, Va., (station No. 02035000) beginning in 1899 and from gaging station No. 02037500 near Richmond, Va., beginning in 1934. Early water-quality data were collected from this station near Richmond, Va., in the early 20th century and published in the Clarke (1924) compilation. Sampling resumed at this station in the 1940s and continued intermittently until 2005. Water-quality data are available from the station at Cartersville beginning in 1929. Several years of water-quality data also are available from the 1940s and 1950s, although regular sampling did not begin until 1968 at this station. Comparing water-quality data from these two stations is complicated by the urban influence around the city of Richmond. However, limiting the use of water-quality data for trend analyses to the Cartersville station provides adequate data and avoids the potential influence of Richmond in the analysis. Comparing the data collected in 1906–07 near Richmond to the more recent data collected upstream of Richmond is expected to result in a conservative estimate of the changes in water quality because of the influence of the city of Richmond on water quality.

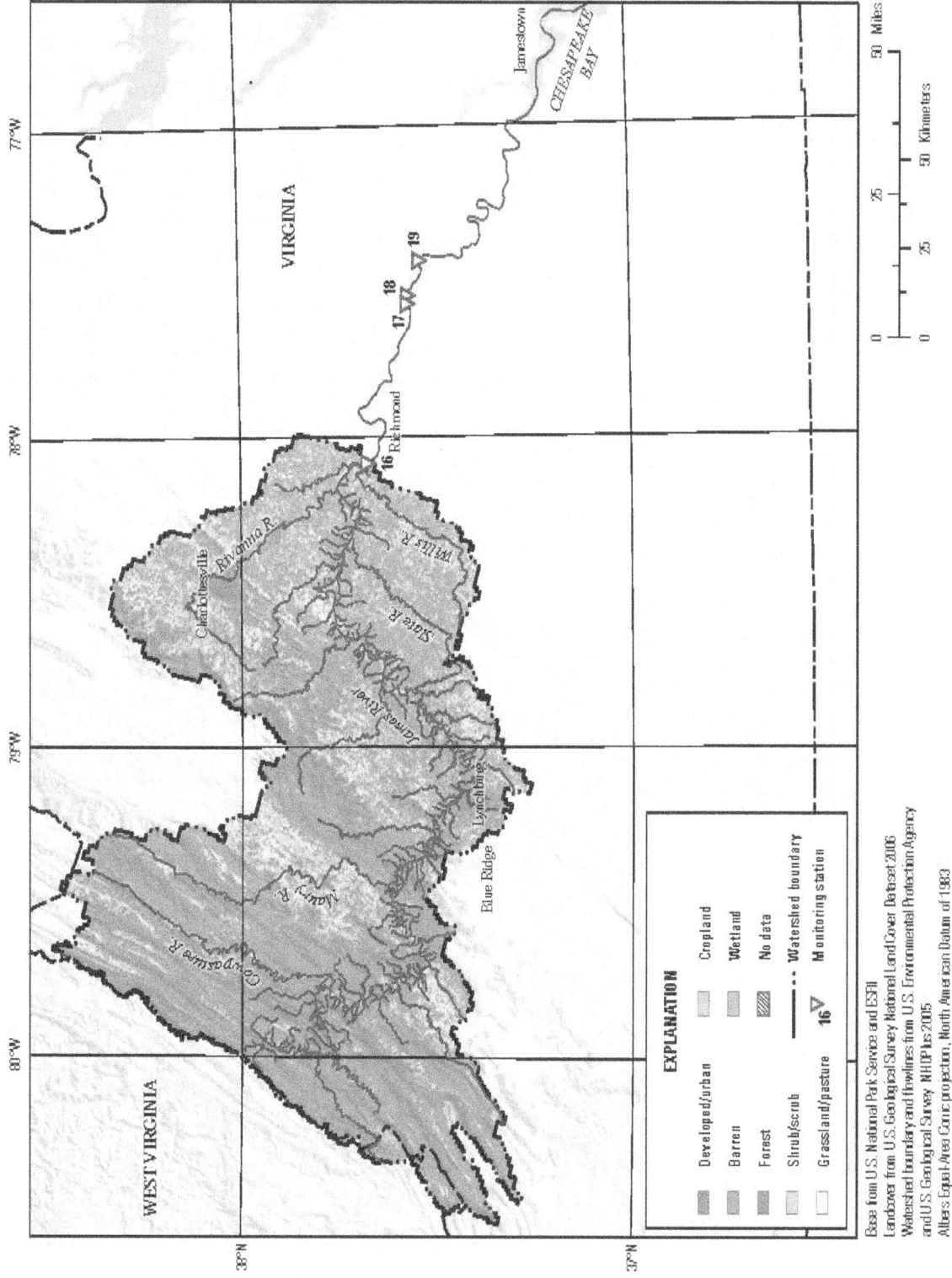

Figure 12. James River basin.

Table 8.   James River basin data sources.

[**Monitoring station** is shown in figure 12. **Source:** NWIS, National Water Information System. **Constituent:** Alk, alkalinity. See table 2 for definitions of all other constituent abbreviations]

Monitoring station	Source	Station	Latitude	Longitude	Constituent	Start year	End year
[1]16	NWIS	02035000 James River at Cartersville, Va.	37°40'15"	78°05'10"	Daily streamflow	1899	2008
					TN	1973	2008
					NH_3	1974	2008
					$TON + NH_3$	1973	1996
					NO_3	1929	2008
					Alk	1929	2008
					TP	1973	2008
					TDP	1973	2008
17	NWIS	02037000 James River and Kanawha Canal near Richmond, Va.	37°33'52"	77°34'28"	Daily streamflow	1936	2008
					NO_3	1952	1972
					Alk	1952	1972
18	NWIS	02037500 James River near Richmond, Va.	37°33'47"	77°32'50"	Daily streamflow	1934	2008
					TN	2004	2005
					NH_3	2004	2005
					NO_3	1906	2005
					Alk	1906	1969
					TP	2004	2005
					TDP	2004	2005
19	NWIS	02037700 James River at Richmond, Va.	37°31'55"	77°26'05"	TN	1979	1981
					$TON + NH_3$	1979	1981
					NO_3	1979	1981
					TP	1979	1981

[1]Reference stream gaging station.

Great Lakes Basins

The size and scope of the Laurentian Great Lakes is difficult to overstate. They cover 244,000 km^2 (94,000 mi^2) and hold 18 percent of the world's freshwater. Theoretical water residence time within the entire basin is well over 100 years. Residence times for individual lakes range from 2.6 years in Lake Erie to almost 200 years in Lake Superior (Fuller and others, 1995). A shorter water residence time indicates higher water input relative to the size of the lake and therefore greater susceptibility of the lake to changes in the quality of the influent water.

The Laurentian Great Lakes system is geologically very young, only about 10,000 years old (Fuller and others, 1995). It was formed by the combined action of a series of glacial advances and retreats associated with recent ice ages. The northern and western parts of the basin are highly influenced by the presence of the Canadian Shield including the Superior Uplands physiographic province (fig. 2), which is characterized by thin, acidic, poorly developed soils. The southern and eastern parts of the basin cover parts of the Central Lowlands physiographic province (fig. 2) and tend to have greater agricultural development and human population densities. The far eastern part of the basin covers parts of the Appalachian Highlands physiographic division, which is moderately developed. As a consequence, human development and its associated effects tend to be concentrated most heavily in the central and eastern parts of the basin. The western part of the basin, particularly around Lake Superior, is only lightly developed.

In this region, two highly contrasting river reaches were considered, the Maumee River at Waterville, Ohio, and the Saint Lawrence River at the outlet of the Great Lakes (fig. 1). The Maumee River is among the most highly agricultural drainage basins included in this study with more than 80 percent of the basin in farmland in 2002 (fig. 4). Fertilizer and animal livestock sources of nitrogen and phosphorus are 7.6 g (N/m^2)/yr and 1.5 g (P/m^2)/yr (figs. 5A–5B). In contrast, less than 20 percent of the Saint Lawrence River basin is in agricultural usage (fig. 4) and nitrogen and phosphorous loading have been low (figs. 5A–5B). Population density in the two basins has been remarkably similar throughout the 20th century (fig. 13A).

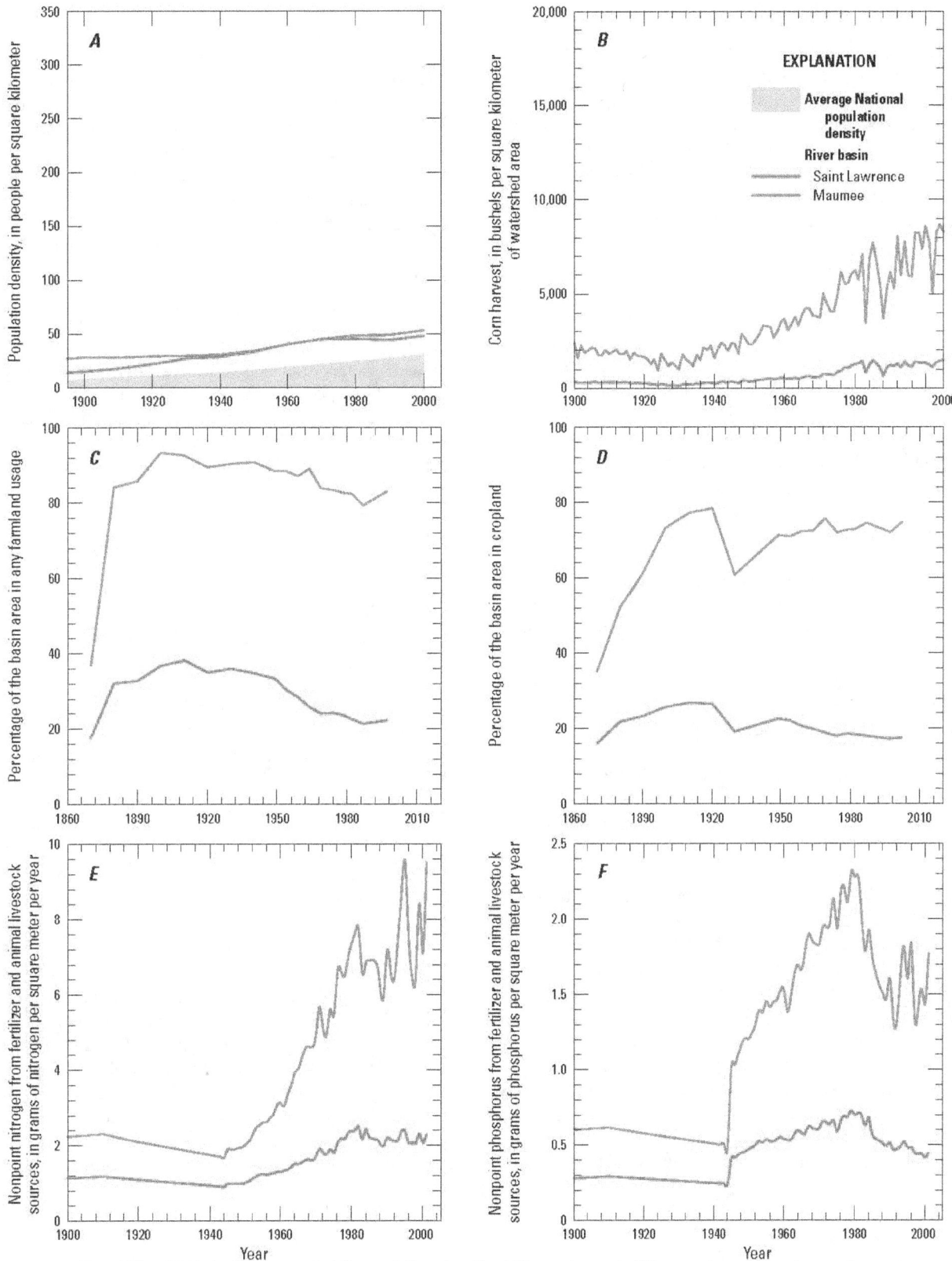

Figure 13. Historical changes in (*A*) population density, (*B*) corn harvest, (*C*) percentage of basin area in any farmland usage, (*D*) percentage of basin area in cropland, and fertilizer and animal livestock sources of (*E*) nitrogen, and (*F*) phosphorus in the Great Lakes basins.

Saint Lawrence River

The Saint Lawrence River has the second highest discharge of any river draining the conterminous United States. The main stem is less than 1,000 km long, but it derives an enormous proportion of its water and drainage area from the Laurentian Great Lakes, which cover over one-half of the basin (fig. 14). The USGS monitors discharge at the Lake Ontario water-level regulation point near Massena, N.Y (the reference stream gaging station, station No. 04264331). There is no stream gage at this location, but discharge is calculated as the sum of several water regulation structures in the area. Therefore, this site is best thought of as the outlet of the Laurentian Great Lakes. At this location, the drainage basin is 774,000 km² (about 300,000 mi²), of which nearly 450,000 km² (173,000 mi²) is within the United States. Mean annual discharge from 1935 to 2009 at this station was approximately 7,100 m³/s (252,000 ft³/s) and mean annual runoff was approximately 291 mm/yr. This station has remarkable temporal stability in streamflow owing to the enormous hydrologic buffering capacity of the Great Lakes as well as the presence of the Lake Ontario water-level regulation structures. Between 1936 and 2008, the entire range of average annual discharges only deviated by about 25 percent from the mean annual discharge, approximately 5,400–8,750 m³/s (200,000–300,000 ft³/s).

The Great Lakes region has long been a focal point of industrial activity and a human population center. In 1990, the combined United States and Canadian human population in the basin was more than 33 million (Fuller and others, 1995). Settlers were originally attracted to the Great Lakes region because of its abundant natural resources including fur-bearing animals, fertile soils, fisheries, mineable ores, and abundant desirable timber resources. More recently, the availability of fresh water and shipping access has encouraged the development of chemical, manufacturing, shipping, and steel industries in this region. Each of these activities has dramatically shaped this region both economically and ecologically. Despite the enormous size of the Laurentian Great Lakes, the combined effects of these human endeavors has substantially altered the Great Lakes ecosystem to the extent that it became severely degraded by the middle of the 20th century.

Impairment of the Great Lakes arguably began in the late 19th century, but culminated in the second half of the 20th century as Lake Erie began to experience frequent hypoxia and noxious algal blooms (Litke, 1999). Problems with water quality were identified as early as 1849 when Chicago began experiencing recurring typhoid fever and cholera outbreaks, which were later linked to contamination of its drinking water source, Lake Michigan (Hill, 2000). In response, Chicago reversed the flow of its primary sanitary sewer, the Chicago River, such that waste from the city would flow into the Des Plaines River and eventually the Illinois River. Ecological disruptions to the Great Lakes continued and worsened throughout the 20th century. The Great Lakes have been the setting for several disastrous ecological invasions that dramatically altered the entire ecosystem and severely damaged native fisheries. The first documented invasion was that of the sea lamprey. Although the lamprey had been observed in Lake Ontario as early as 1830, it only began spreading to the other Great Lakes after improvements to the Welland Canal in 1919, which allowed the lamprey to enter Lake Erie (Great Lakes Fishery Commission, 2000). By the 1950s, the lamprey had spread to all of the Great Lakes and greatly reduced populations of several valuable native species, such as lake trout, whitefish, and chub (Great Lakes Fishery Commission, 2000). The alewife was the next important invasion. The alewife arrived in the Great Lakes by the late 1940s. This fish is native to the Atlantic Ocean, but is able to survive in the Great Lakes and has grown large populations in Lake Michigan and Lake Huron. The alewife primarily feeds on zooplankton placing it in direct competition with native lake herring, whitefish, chub, and perch (University of Wisconsin Sea Grant, 2002), further stressing these native fish. More recently, zebra mussel invasions of the Great Lakes also have had complex and far-ranging effects on the ecosystem. Zebra mussels are filter feeders and have greatly increased water clarity. Some native species benefit, such as northern pike and yellow perch, but zebra mussels displaced the native unionid clams in Lake Erie and Lake Saint Clair (U.S. Geological Survey, 2011).

Pollution inputs to the Great Lakes also increased throughout the first half of the 20th century and resulted in water-quality degradation of the lakes and many of the lake tributaries. Eutrophication was a severe problem in the Great Lakes by the mid-20th century, particularly in Lake Erie (Conference on Water Pollution and the Great Lakes, 1961). Industrial pollutants are a continuing concern in the Great Lakes, including mercury, legacy DDT, and polychlorinated biphenyls. Although the input of these pollutants has greatly decreased in recent decades, fish consumption advisories are still in place for many fish species (Great Lakes Information Network, 2012). In 1969, a fire on the Cuyahoga River, which had burned repeatedly throughout the 20th century, crystallized national discontent over the rapidly deteriorating quality of waters throughout the Nation and helped provide the necessary impetus for passage of the Clean Water Act (Melosi, 2000).

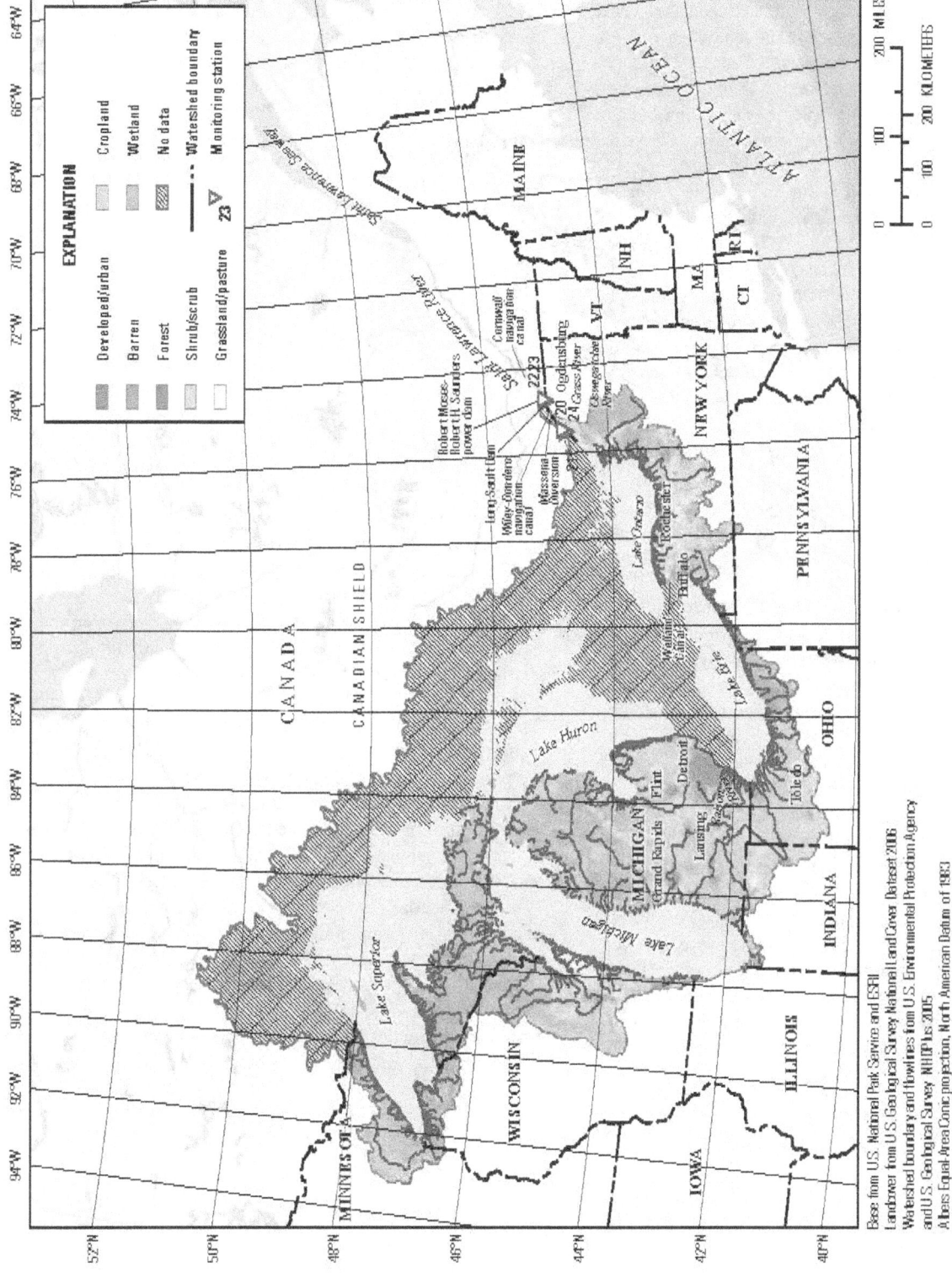

Figure 14. Saint Lawrence River basin.

Additional attention has been focused on remediation and protection of the Great Lakes through the Great Lakes Water Quality Agreement, which is a lake management agreement between entities in the United States and Canada. The agreement was signed in 1978 and emphasizes an ecosystem-based approach to address the problems confronting the Great Lakes (Fuller and others, 1995). Implementation of the Clean Water Act and the Great Lakes Water Quality Agreement resulted in major reductions in pollution including reductions in toxic pollutant inputs, phosphorus inputs, and floating debris (Fuller and others, 1995). In 1987, updates to the Great Lakes Water Quality Agreement also included designations for areas of concern with special remedial action plans to be implemented to restore beneficial uses to the Great Lakes.

In the United States part of the basin, human population density was 48 people/km² in 2000, which was greater than the national average of 31 people/km² and placed this basin approximately in the middle of basins considered in this study (fig. 3). Fuller and others (1995) reported a similar population density for the entire basin, approximately 43 people/km² (including the Canadian portion of the basin). In 2002, farmland covered 22 percent of the entire basin and cropland covered 17 percent (fig. 4), but there are strong gradients in farmland coverage such that agricultural land use varies from 6 percent surrounding Lake Superior to 60 percent around Lake Erie (Fuller and others, 1995). In the United States portion of the basin, farmland area peaked early in the 20th century with nearly 40 percent of the basin in agricultural land usage. However, the total amount of farmland has decreased steadily since the 1940s. In the parts of the basin downstream of the Great Lakes, forest cover increased from 37 to 60 percent over the latter half of the 20th century (Benke and Cushing, 2005). Fertilizer and animal livestock sources of nitrogen are relatively low with animal livestock comprising a very small portion of total inputs. On average from 1997 to 2001, fertilizer and animal livestock sources of nitrogen were 0.6 and 0.3 g (N/m²)/yr (fig. 5A). Phosphorus sources from fertilizers and animal livestock were similar, each averaging approximately 0.1 g (P/m²)/yr between 1997 and 2001 (fig. 5B). It is important to recognize that point sources of phosphorus were a severe problem in the Great Lakes in the middle and late 20th century and were a major focus of an intense effort to clean up the Great Lakes (Litke, 1999). Average corn harvest for the United States part of the basin during 2000–2010 was 1,464 bushels/km² of basin area (fig. 6), the eighth highest for all basins considered in this study.

Data Sources

Data sources for the Saint Lawrence River are summarized in table 9. The USGS has monitored discharge from Lake Ontario at Cornwall, Ontario near Massena, N.Y., since 1935 (reference stream gaging station, station No. 04264331). However, there is no stream gage at this location. Instead, discharge is determined from summation of discharge through the Robert Moses-Robert H. Saunders power dam, the Long Sault Dam, the Massena Diversion, the Raisin River Diversion, the Cornwall and Massena municipal water supply, and the Cornwall and the Wiley-Dondero navigation canals. Nitrate and alkalinity are available at the reference stream gaging station near Massena, N.Y. since 1955 although regular sampling did not begin until 1974 (table 9). The Saint Lawrence River also was included in the Clarke (1924) publication with 11 discrete samples having been taken in 1906 and 1907 at Ogdensburg, N.Y. Ogdensburg is upstream of Cornwall and misses inputs from several tributaries including the Oswegatchie and Grass Rivers. However, the basin area is reported to be 765,000 km² (295,000 mi²) and Clarke (1924) reports that discharge at this location was 248,000 ft³/s (7,000 m³/s). Both of these properties are within 2 percent of the calculation for the reference stream gaging station and so data from the Ogendsburg and Cornwall stations are presumed to be adequately representative of water quality in this part of the Saint Lawrence River. Additional water-quality sampling was performed at the Saint Lawrence River at Cornwall, Ontario, by the Canada Department of Mines and Resources, and published in a series called "Industrial Water of Canada" (Leverin, 1947). These include annual nitrate, alkalinity, and major ion observations from 1935 to 1940, which were reported in USGS Water Supply Paper 2400 along with a series of discrete samples taken from 1947 to 1948 (table 9; U.S. Geological Survey, 1993). The New York State Department of Health also reported several water-quality observations taken from Ogdensburg, N.Y., in 1914 (table 9).

Table 9. Saint Lawrence River basin data sources.

[**Monitoring station** is shown in figure 14. **Source:** NWIS, National Water Information System; USGS, U.S. Geological Survey; WSP, Water-Supply Paper. **Constituent:** Alk, alkalinity. See table 2 for definitions of all other constituent abbreviations]

Monitoring station	Source	Station	Latitude	Longitude	Constituent	Start year	End year
[1]20	NWIS	04264331 Saint Lawrence River at Cornwall, Ontario, near Massena, N.Y.	45°00′22″	74°47′42″	Daily streamflow	1935	2009
					TN	1974	2009
					NH_3	1977	2009
					$TON+NH_3$	1974	2009
					NO_3	1955	2009
					Alk	1955	2009
					TP	1974	2009
					TDP	1977	2009
21	NWIS	04264000 Saint Lawrence River at Ogdensburg, N.Y.	44°41′20″	75°30′45″	TN	1968	1971
					NH_3	1968	1971
					$TON+NH_3$	1968	1971
					NO_3	1906	1971
					Alk	1906	1971
					TP	1966	1971
[2]22	USGS WSP 2400	Saint Lawrence River at Cornwall, Ontario	45°00′22″	74°47′42″	NO_3	1935	1940
					Alk	1935	1940
[2]23	Canada Department of Mines and Resources	Saint Lawrence River at Cornwall, Ontario	45°00′22″	74°47′42″	NO_3	1947	1948
					Alk	1947	1948
[2]24	New York State Department of Health	Saint Lawrence River at Ogdensburg, N.Y.	44°41′20″	75°30′45″	NH_3	1914	1914
					NO_3	1914	1914
					Alk	1914	1914

[1]Reference stream gaging station.

[2]Historical station, location approximate.

Maumee River

The Maumee River is located in the Great Lakes basin and covers portions of Michigan, Ohio, and Indiana (fig. 15). It is the largest river basin draining to the Great Lakes, about 17,000 km² (6,600 mi²). The Maumee River is the largest single source of sediment to the Great Lakes (Myers and others, 2000). Mean annual discharge at the reference stream gaging station (station No. 04193500) for the period 1898–2008 is 150 m³/s (5,300 ft³/s) and runoff is 280 mm/yr. The Maumee River is formed by the confluence of the Saint Joseph and Saint Mary Rivers in Fort Wayne, Ind., and flows approximately 130 mi before emptying into Lake Erie at Toledo, Ohio. The entire basin is located within the Central Lowlands physiographic province (fig. 2), which was originally covered with temperate deciduous forest, but has been extensively cultivated.

More than 90 percent of the soils in the basin are poorly to very poorly drained. The central basin was originally covered with a large wetland called the Black Swamp. Agricultural development of this area required draining the wetland through a network of ditches and tile drains (Myers and others, 2000). Poor water retention and the presence of these drainage structures in the basin contribute to runoff and carry sediments from drained agricultural areas to stream channels and other water bodies.

Water-quality problems in the Maumee River basin relating to agricultural nonpoint pollution and other pollution sources prompted the Environmental Protection Agency to list the basin as one of its "Areas of Concern" for the Great Lakes region (U.S. Environmental Protection Agency, 2006). The increase in suspended solid load by the river is of particular concern. Soil erosion in the Maumee River basin increases sediment transport from the upper parts of the basin toward the mouth. The increased sediment load has negative economic, ecological, and human health effects. Fish and aquatic invertebrate communities often suffer from significant changes in sediment load. Accordingly, some of the most highly altered fish and aquatic invertebrate communities in Ohio reside within the Maumee River basin (Myers and others, 2000). High sediment loads also require increased dredging to maintain waterways including the port of Toledo, Maumee Bay, and Federal waterways along the lower river channel. Efforts to control sedimentation in this basin include adoption of conservation tillage practices by farms in the basin. By the late 1990s, nearly one-half of all farm acres were in conservation tillage (Myers and others, 2000).

The Port of Toledo is an important shipping node in the Laurentian Great Lakes, which greatly contributed to the Maumee River basin becoming an important regional manufacturing hub. Pollution associated with manufacturing is a continuing concern in the Maumee River, including PCB (polychlorinated biphenyl) and mercury contamination in the lower river segments (Benke and Cushing, 2005). The presence of these contaminants exacerbates the sediment pollution problem because it complicates disposal of the dredge material (Myers and others, 2000).

The largest population centers in the basin are Toledo, Ohio, located at the mouth of the river, and Fort Wayne, Ind., at the headwaters. Overall population density in the basin in 2000 was 53 people/km², which is higher than the national average of 31 people/km² and makes the Maumee River basin the eighth most populous basin included in this study (fig. 3; U.S. Census Bureau, 2000). The city of Toledo built its first coordinated sewerage system, called the Bay View facility, located near the mouth of the Maumee River, in 1922 although primary wastewater treatment did not begin until 1932 (City of Toledo, 2010). Secondary treatment of wastewater began in 1959 along with grit and grease removal. Capacity of the wastewater-treatment facility was increased several times in the second half of the 20th century. Combined sewer overflows remain a concern in older parts of the system and in 1983, the city built a facility to provide settling treatment to storm water overflows.

The basin has a very high degree of agricultural development with 75 percent of the basin in cropland and another 8 percent in other agricultural uses (fig. 4). Agricultural development occurred later in this drainage basin than several other Midwestern streams included in this study. Maximum cropland area, more than 78 percent, was recorded in the 1920 Census of Agriculture (fig. 13; M. Haines, Colgate University, written commun., 2004) whereas other heavily agricultural basins, such as the Des Moines and Illinois River, reached maximum cropland area prior to 1900. Nevertheless, the Maumee River basin remains among with most highly developed agricultural basins in this study. Nitrogen fertilizer application rates averaged 6.5 g (N/m²)/yr between 1997 and 2001 and were comparable to the Illinois River basin (fig. 5). Animal livestock sources accounted for another 1.1 g (N/m²)/yr (fig. 5). Similarly, fertilizer and animal livestock sources of phosphorus were among the highest in this study, averaging more than 1.5 g (P/m²)/yr from 1997 to 2001 (fig. 5). This basin also has the highest output of corn outside of the upper Mississippi River basin, averaging more than 8,000 bushels/km² of drainage area between 2000 and 2010 (fig. 6). The Maumee River basin exemplifies the tremendous increases in fertilizer usage and corn production that occurred in the second half of the 20th century. Fertilizer and animal livestock sources of nitrogen were around 2 g (N/m²)/yr and then increased dramatically in the latter decades of the 20th century (fig. 13E). Phosphorus sources show a similar pattern, but there is a notable decrease from about 2.3 to 1.7 g (P/m²)/yr between 1980 and 2000 (fig. 13E).

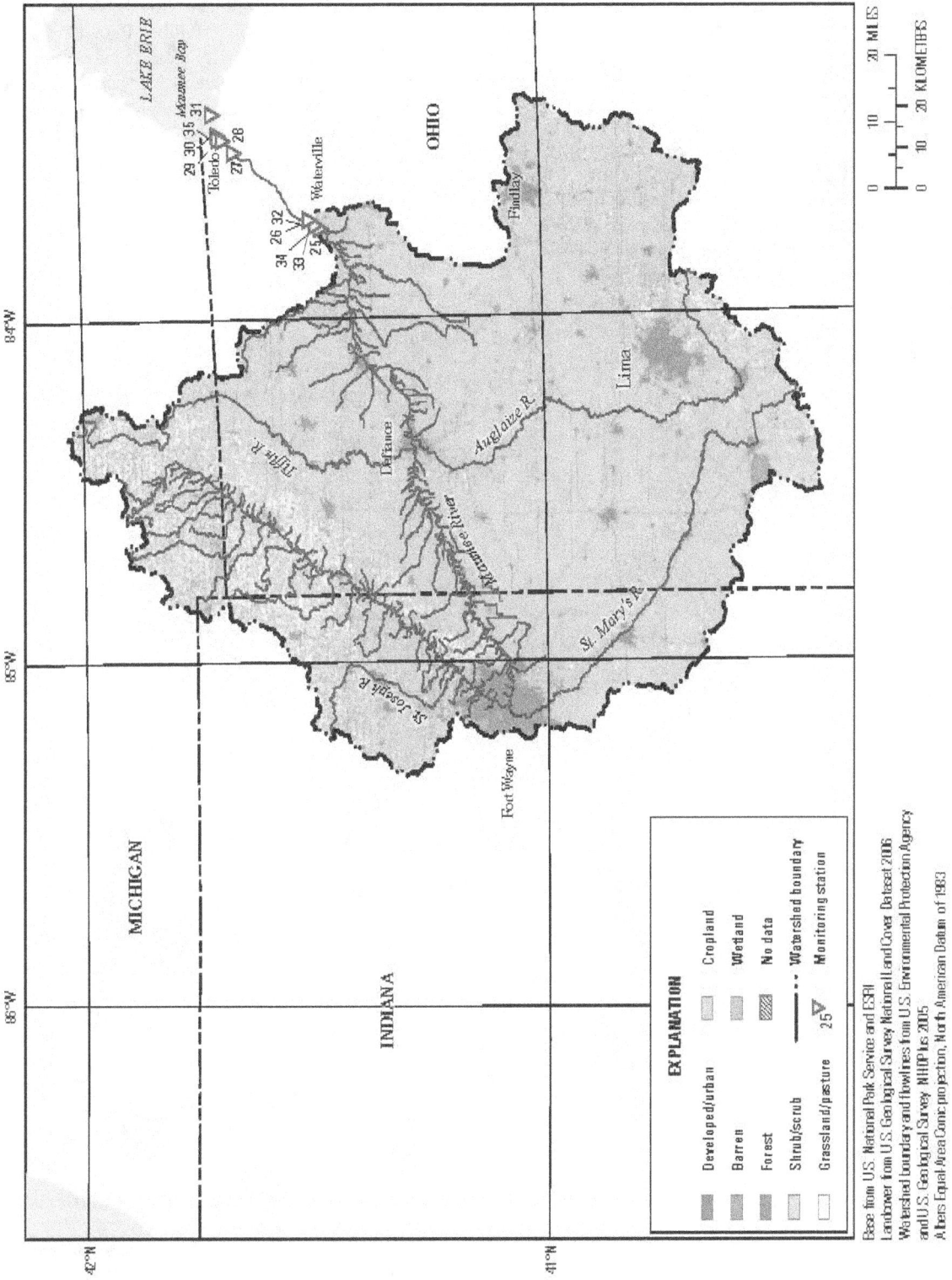

Figure 15. Maumee River basin.

Base from U.S. National Park Service and ESRI
Landcover from U.S. Geological Survey National Land Cover Dataset 2006
Watershed boundary and flowlines from U.S. Environmental Protection Agency
and U.S. Geological Survey NHDPlus 2005
Albers Equal Area Conic projection, North American Datum of 1983

Data Sources

Data sources for the Maumee River are summarized in table 10. Daily stream gage data began to be collected in 1898 at the reference stream gaging station at Waterville, Ohio, (station No. 04193500) with a gap from January 1936 to March 1939. The basin area at this gage, is 16,400 km² (6,330 mi²) or about 10 percent less than at the mouth of the river. However, this station has the longest stream-gaging record on this segment of the river. Substantial amounts of water-quality data are available at Waterville from the USGS and the Ohio Environmental Protection Agency for the second half of the 20th century. Additional water-quality data, including samples collected in 1906–07, are available from sampling stations located in Toledo, Ohio. Notably, these stations are downstream of the Bay View wastewater-treatment facility and so water quality will be affected by its effluent. During 1950–66, water-quality data are available only from near the mouth of the Maumee River and so provide the only frame of reference. Since 1965, data have become available at Waterville, Ohio. Inferences drawn from water-quality trends in this basin must take these considerations into account. It is presumed that the water-quality data collected near the mouth of the Maumee River during 1950–66 were affected by the effluent of the wastewater-treatment plant, including elevated nutrient concentrations. More recent data collected at Waterville, Ohio, about 15 mi upstream of Toledo, was not influenced by wastewater from Toledo and so the inferred trends from the data could be conservative.

Table 10.   Maumee River basin data sources.

[**Monitoring station** is shown in figure 15. **Source:** NWIS, National Water Information System; STORET, Storage and Retrieval Data Warehouse. **Constituent:** Alk, alkalinity. See table 2 for definitions of all other constituent abbreviations]

Monitoring station	Source	Station	Latitude	Longitude	Constituent	Start year	End year
25	NWIS	04193490 Maumee River near Waterville, Ohio	41°28'34"	83°44'20"	TN NH_3 $TON + NH_3$ NO_3 TP	1979 1979 1979 1979 1979	1989 1980 1989 1989 1989
[1]26	NWIS	04193500 Maumee River at Waterville, Ohio	41°30'00"	83°42'46"	Daily streamflow TN NH_3 $TON + NH_3$ NO_3 Alk TP TDP	1898 1974 1977 1974 1966 1965 1968 1977	2008 2007 2007 2007 2007 2007 2007 2007
27	NWIS	04194010 Maumee River at Craig Bridge, at Toledo, Ohio	41°39'46"	83°30'29"	NO_3 Alk TP	1962 1962 1962	1966 1966 1966
28	NWIS	04194022 Maumee River at Toledo OSEAS Terminal Dock, Ohio	41°41'06"	83°28'35"	NO_3 Alk TP	1962 1962 1962	1974 1974 1974
29	NWIS; Clarke, 1924	04194023 Maumee River at mouth, at Toledo, Ohio	41°41'36"	83°28'20"	NH_3 NO_3 Alk TP	1974 1906 1906 1974	1975 1975 1975 1975
30	NWIS	04194030 Maumee River at C & O Dock, at Toledo, Ohio	41°41'46"	83°27'39"	NO_3 Alk TP	1962 1962 1962	1974 1974 1974
31	NWIS	04194050 Maumee River at Buoy 39, at Toledo, Ohio	41°42'22"	83°24'0"	NO_3 Alk TP	1962 1962 1962	1966 1966 1966
32	STORET	500080 Maumee River at Waterville, Ohio	41°30'00"	83°42'46"	TN NH_3 $TON + NH_3$ NO_3 Alk TP TDP	1973 1973 1973 1973 1973 1968 1974	1998 1998 1998 1998 1998 1998 1996
33	Heidelberg University	USGS04193500 Maumee River at Waterville, Ohio	41°29'12"	83°42'46"	TN NH_3 $TON + NH_3$ NO_3 Alk TP	1976 1975 1976 1975 1993 1975	1998 1998 1998 1998 1997 1998
[2]34	Love, 1954, 1955, 1956a	Maumee River at Waterville, Ohio	41°29'12"	83°42'46"	NO_3 Alk	1950 1950	1952 1952
[2]35	Love, 1954, 1955, 1956a	Maumee River at Toledo OSEAS Dock, Ohio	41°41'06"	83°27'39"	NO_3 Alk	1950 1950	1952 1952

[1]Reference stream gaging station.

[2]Historical station, location approximate.

Central Basins

Central basins include the Mississippi River basin, which incorporates the Ohio River basin in the east, the Missouri River basin in the west, as well as the upper, middle, and lower Mississippi River. The Illinois River, Des Moines River, and Arkansas River are other major tributaries included in this study (fig. 1). The Brazos River also was included in this category because of its similarity and geographic proximity to the other river basins west of the Mississippi River main stem, such as the Arkansas and the Missouri. The Brazos River basin is part of the Texas Gulf Hydrologic region.

The Mississippi River is the largest river in North America with a drainage basin covering 3.2 million km^2 (1.2 million mi^2), about 40 percent of the continental United States. The Mississippi River, together with its main distributary, the Atchafalaya River, discharges about 800,000 ft^3/s into the northern Gulf of Mexico. This is almost 40 percent of the total river flow from the continental United States to the coastal ocean. Substantial hydrologic modification of the basin has occurred throughout the main stem Mississippi through the construction of wing-dams, levees, and lock-and-dam structures designed to maintain a navigable channel. Major tributaries of the Mississippi River including the Ohio, Illinois, Missouri, and Arkansas Rivers also have been extensively dammed for navigation, municipal water supply, irrigation, and flood control projects.

All or part of 31 States and two Canadian Provinces reside within the Mississippi River basin, making a broad description of the basin difficult. Accordingly, the factors influencing the water quality in the main stem Mississippi River are certainly diverse. Intense agriculture occurs throughout much of the basin, particularly in the Upper Midwestern United States and along the alluvial floodplain of the lower Mississippi River. Major population centers, such as Chicago, Ill., Nashville, Tenn., St. Louis, Mo., Minneapolis-St. Paul, Minn., and Denver, Colo., are located within the drainage basin boundaries. Because of the large geographic extent of this region and the number of river reaches of interest, the discussion of ancillary data is focused into three geographic regions: basins east of the main stem of the Mississippi River, basins west of the main stem of the Mississippi River, and the Mississippi River main stem itself.

River reaches of interest to this study in the eastern part of the Central region include the Middle and Lower Illinois River and the Middle and Lower Ohio River (fig. 1). The Illinois River is among the most densely populated, and most highly agricultural basins in this study (figs. 3 and 4). In contrast, the Ohio River has much lower population density and a much lower percentage of the basin in farmland (figs. 3 and 4). The Illinois River basin is most affected by large urban centers, encompassing densely populated areas of northern and central Illinois. This basin has been subject to relatively high population density for most of the 20th century, with population density reaching 100 people/km^2 by 1920 (fig. 16A). By contrast, population density within the Ohio River basin was less than 100 people/km^2 by the end of the 20th century (fig. 16A). Middle and Lower Illinois River basins are 70 and 80 percent farmland, respectively, and major agricultural development occurred before 1900 (fig. 16C). The Ohio River basin reached 80 percent farmland by 1900 and has steadily decreased throughout the 20th century to around 40 percent in 2002 (fig. 16C). The Lower and Middle Illinois River reaches also had among the highest average corn harvest rates from 2000 to 2010 (fig. 6), and among the highest intensity of nitrogen sources from fertilizer and animal livestock (figs. 5A–5B).

River reaches west of the Mississippi River include the Missouri, Arkansas, and Brazos Rivers, which are very large basins in the more arid Great Plains (fig. 1). The Des Moines River also is west of the Mississippi main stem, but it contrasts strongly with the other basins. All basins have low population density, less than 21 people/km^2 in the year 2000 (fig. 3), and a large proportion of the basin in farmland (fig. 4). However, the Des Moines River basin is more than 75 percent cropland while the Missouri, Arkansas, and Brazos River basins each have less than 40 percent of the basin in cropland (fig. 4). Of all the basins considered in this study, the Des Moines River has the highest identified sources of nitrogen and phosphorus from fertilizer and animal livestock (figs. 5A–5B). It also averaged the second highest corn harvest rate from 2000 to 2010 (fig. 6). In contrast, the Missouri, Arkansas, and Brazos Rivers have much lower corn harvests and only a small fraction of the nitrogen and phosphorus sources from fertilizer and animal livestock (figs. 5A–5B). Farmland

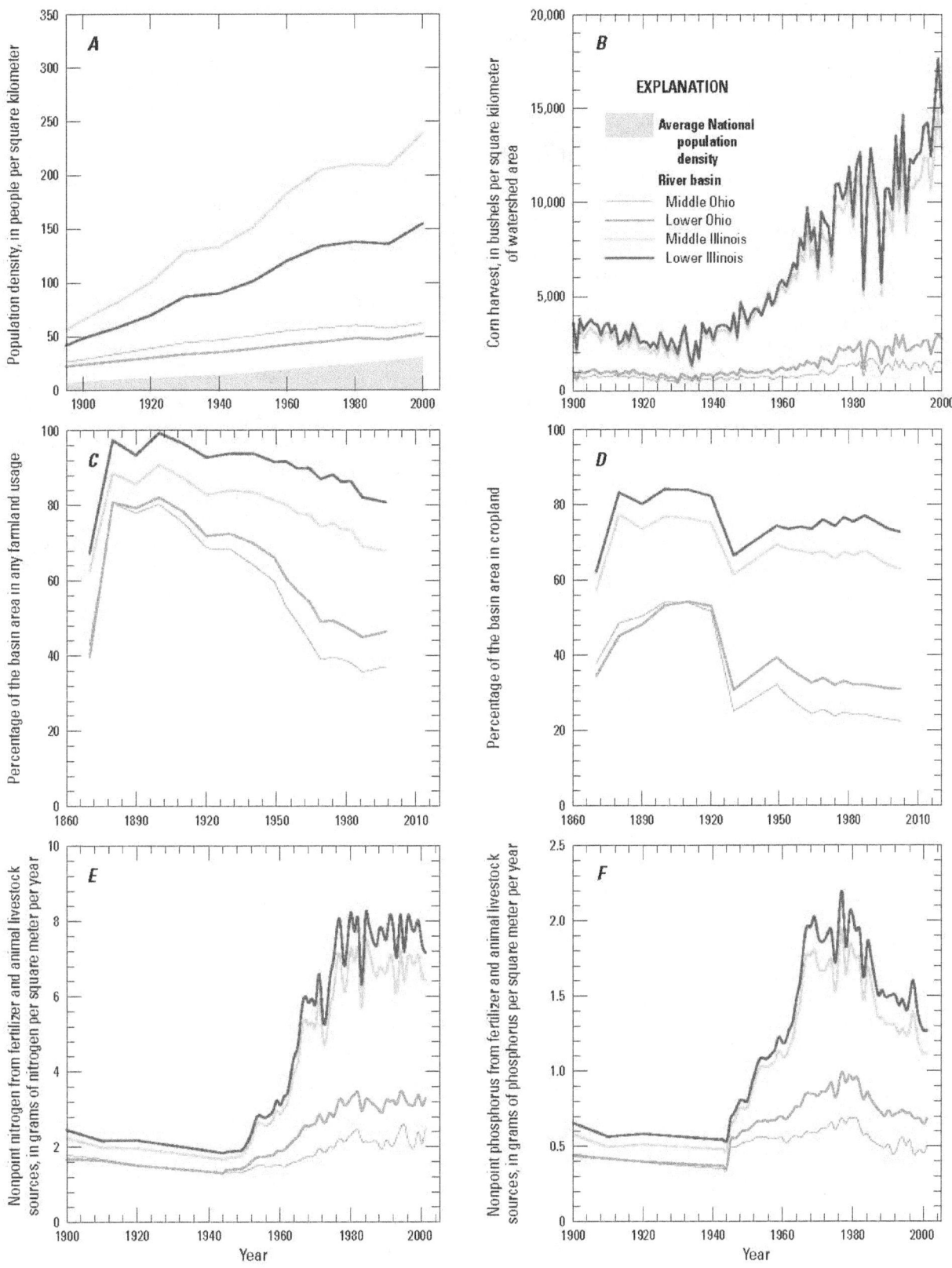

Figure 16.   Historical changes in (A) population density, (B) corn harvest, (C) percentage of basin area in any farmland usage, (D) percentage of basin area in cropland, and fertilizer and animal livestock sources of (E) nitrogen and (F) phosphorus in the Central basins east of the main stem of the Mississippi River.

development in all of these basins west of the Mississippi River proceeded along a similar timeline and trajectory with the basins becoming heavily agricultural by the early 20th century (fig. 17C). However, the Des Moines River basin had approximately double the amount of cropland in the basin as the other basins as early as 1900 (fig. 17D). These data emphasize the contrasting nature of agricultural development and the implications on the identified sources of nitrogen and phosphorus in heavily cultivated river basins, such as the Maumee (figs.13B–13D), Illinois (figs. 16B–16D), and Des Moines Rivers (figs. 17B–17D) as compared with those of the Great Plains (Missouri, Arkansas, and Brazos Rivers) (figs. 17B–17D).

The river reaches of interest along the main stem of the Mississippi River largely reflect regional influences of the basin. Accordingly, the Upper Mississippi River, which includes the Illinois and Des Moines Rivers as well as other parts of eastern Iowa, southern Minnesota, and western Illinois, has a higher population density, corn harvest rate, farmland and cropland percentage, and identified sources of nitrogen and phosphorus in the basin (figs. 18A–18F). In contrast, the Middle and Lower parts of the main stem of the Mississippi River are similar and reflect the influences of several large tributaries including the Missouri, Ohio, and Arkansas Rivers. Although the percentage of basin area

in farmland acreage in these basins is similar to that of the Upper Mississippi (fig. 18C), these basins have much lower cropland percentages (fig. 18D). Corn harvest rates also are much lower (fig. 18B) as are the identified sources of nitrogen and phosphorus from fertilizer and animal livestock (figs. 18E–18F).

Understanding the nutrient dynamics on the main stem Mississippi River is particularly important because of the contribution to the hypoxia that develops annually in the northern Gulf of Mexico. On average, the hypoxic area covers about 16,500 km² (6,400 mi²) and is the second largest coastal hypoxic zone in the world (U.S. Environmental Protection Agency Science Advisory Board, 2007). Coastal hypoxia is a growing problem worldwide and is caused by excess nutrient inputs that foster algal blooms. The death and decomposition of algae following blooms consumes oxygen and causes hypoxic conditions, depending on the stability of the water column and the rate of oxygen consumption. Although there is some evidence for hypoxic events in the northern Gulf of Mexico occurring more than 100 years ago, the intensity and magnitude of these events has increased greatly in the past several decades (U.S. Environmental Protection Agency Science Advisory Board, 2007). It is clearly of interest to investigate long-term patterns in nutrient dynamics in the Mississippi River.

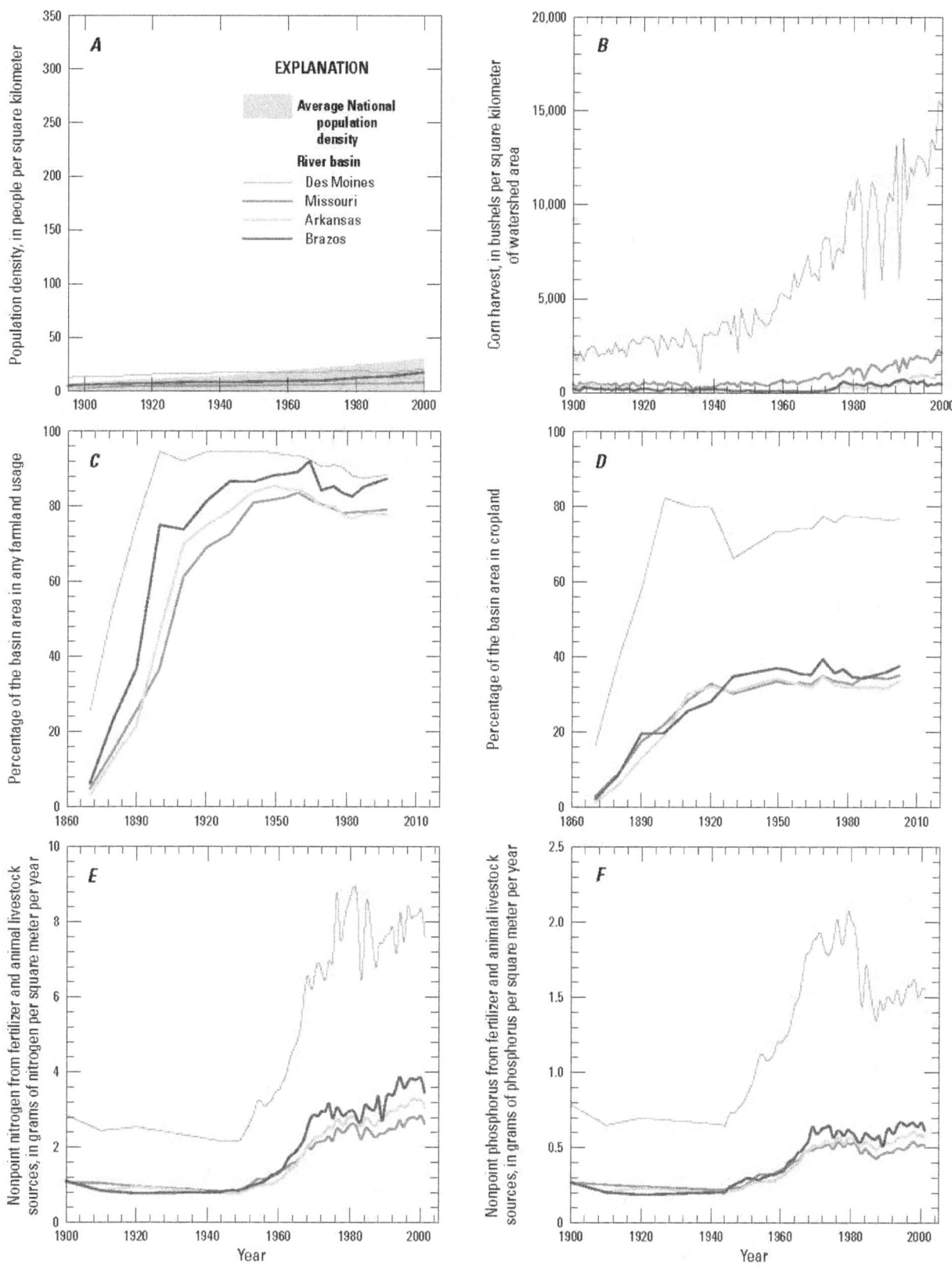

Figure 17.   Historical changes in (A) population density, (B) corn harvest, (C) percentage of basin area in any farmland usage, (D) percentage of basin area in cropland, and fertilizer and animal livestock sources of (E) nitrogen and (F) phosphorus in the Central basins west of the main stem of the Mississippi River.

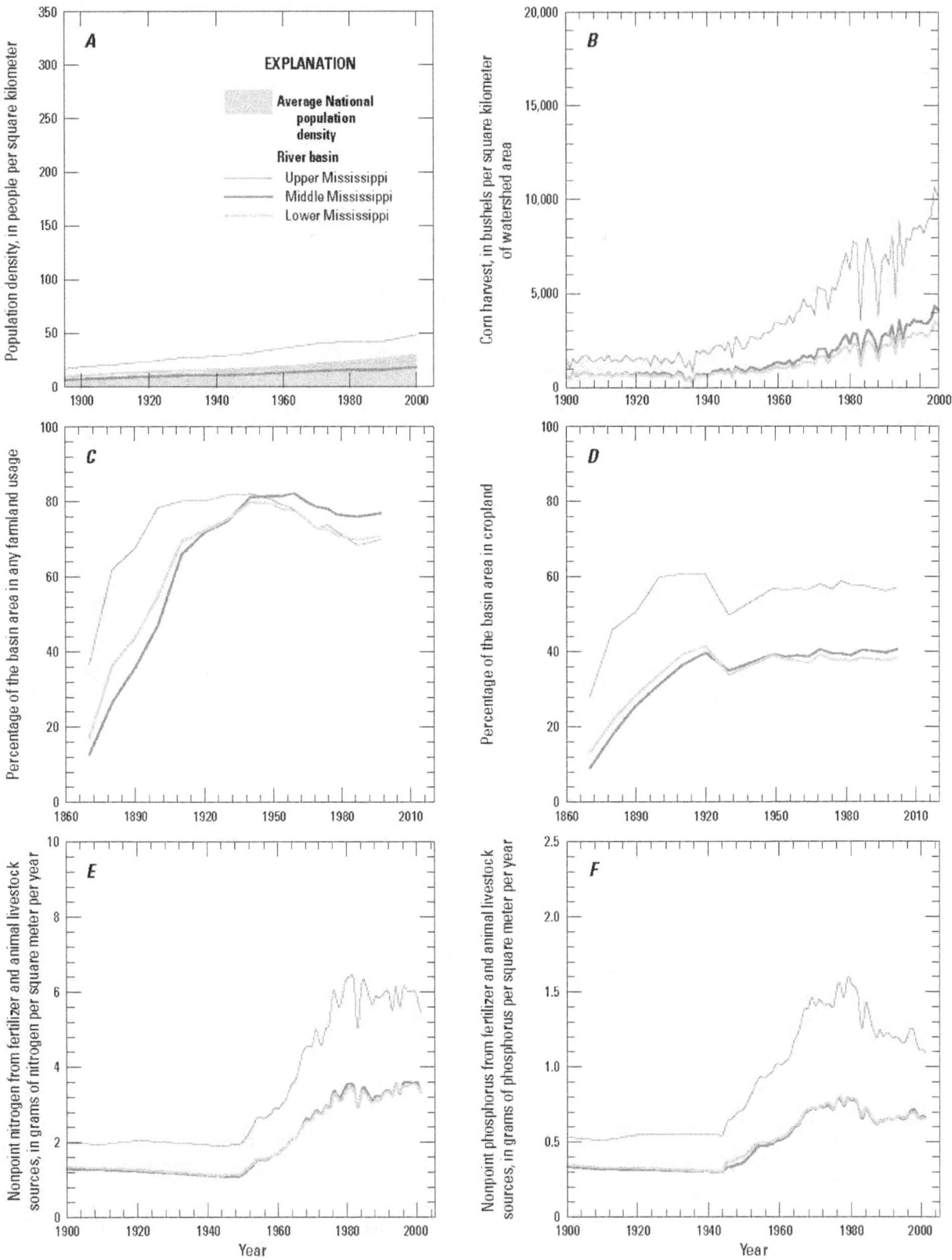

Figure 18. Historical changes in (A) population density, (B) corn harvest, (C) percentage of basin area in any farmland usage, (D) percentage of basin area in cropland, and fertilizer and animal livestock sources of (E) nitrogen, and (F) phosphorus in the basins of interest on the main stem of the Mississippi River.

Ohio River

The Ohio River is the third largest river in terms of discharge in the United States. Measured at the reference stream gaging station for the lower river at Metropolis, Ill. (station No. 03611500), mean annual discharge (1928–2008) is 7,900 m^3/s (278,000 ft^3/s) and runoff is 472 mm/yr. The Ohio River basin has the highest runoff of any major tributary to the Mississippi River and contributes 40 percent of the total discharge of the Mississippi River while making up only 16 percent of the total basin area, approximately 526,000 km^2 (203,000 mi^2). The Ohio River is formed in Pittsburgh, Pa., by the confluence of the Allegheny and Monongahela Rivers (fig. 19). Flow at the confluence of these rivers is about 960 m^3/s (34,000 ft^3/s) and runoff is nearly 600 mm/yr. The river then flows almost 1,600 km to its confluence with the Mississippi at Cairo, Ill. Benke and Cushing (2005) identify 12 major tributaries to the Ohio River, which account for 84 percent of the total basin area. They include the Allegheny, Monongahela, Scioto, Little Miami, Great Miami, Licking, Kentucky, Green, Kanawha, Wabash, Cumberland, and Tennessee Rivers. The major tributaries encompass a wide range of physiographic conditions including steep, rocky streams draining the Appalachian Mountains to gently sloping streams draining deep till plains. The Tennessee and Cumberland River basins also are noted for their high diversity of fishes and aquatic invertebrates (Benke and Cushing, 2005). Flow on the river is tightly controlled for river navigation, with the first dam built on the river almost 200 years ago at Henderson, Ky. (Benke and Cushing, 2005). The Ohio River Valley is highly industrialized and has been at the center of the American industrial economy for over a century. Consequently, the river has a long history of water-quality problems arising from mining, industrial activities, and population centers.

The first coordinated water quality study of the Ohio River was conducted in 1914–1915 by the U.S. Public Health Service. Passage of the Public Health Service Act of 1912 paved the way for Federal studies of water pollution in the United States (Scarpino, 1985). The Ohio River was the first river to be studied by the U.S. Public Health Service. Early in the 20th century, interacting problems of untreated sewage disposal, industrial pollution, and acidic mine drainage resulted in degradation of water quality in the river. Major metropolitan areas of Pittsburgh, Cincinnati, and Louisville discharged considerable amounts of untreated sewage into the river. However, the presence of industrial waste and acidic waters inhibited "self-purification," the usual process of biological remediation of pollution in the river (Hoskins and others, 1927). In the early 20th century, 38 low-head navigational dams were constructed along the Ohio River (Hoskins and others, 1927). The dams created a series of slackwater pools that trapped particulate matter and interrupted downstream transport of polluted material. Flood events often scoured sediments behind the navigational dams and resulted in a pulse of polluted material being transported downstream, which led to undesirable conditions and complicated drinking water treatment (Cleary, 1967).

In the early 1920s, phenol pollution became a prominent issue on the Ohio River. Even in minute amounts, phenols impart a medicinal aroma to drinking water, a quality that is exacerbated by chlorination (Cleary, 1967). State and municipal officials came to an informal agreement with industry leaders to minimize phenol discharges into the Ohio River and its tributaries, although there was no legal mechanism to enforce this agreement (Cleary, 1967). The nature of this agreement exemplifies the difficulties encountered by public health officials trying to coordinate water-quality improvements in the Ohio River.

As previously discussed, efforts to strengthen the Federal role in guaranteeing water quality in the Nation's rivers proceeded very slowly throughout the 20th century. Although the preferred model was one in which the States took lead responsibility for cleaning up their own waters, the Ohio River provided a counter example of a region in urgent need of water-quality improvement that no single State could achieve on its own. As a result, by the middle of the century eight States bordering the main stem Ohio River joined together to form a regional compact known as the Ohio River Valley Water Sanitation Commission (ORSANCO). Water sanitation officials and industry leaders began discussing the idea in 1935, but the final commission did not form until 1948 (Cleary, 1967).

The objective of ORSANCO was to control existing and future pollution in rivers and streams in the Ohio River basin (Cleary, 1967). ORSANCO relied upon the cooperative nature of the commission in setting out to achieve these goals. By the mid-1960s, almost $1 billion had been spent on cleanup and wastewater treatment in the Ohio River basin (Cleary, 1967).

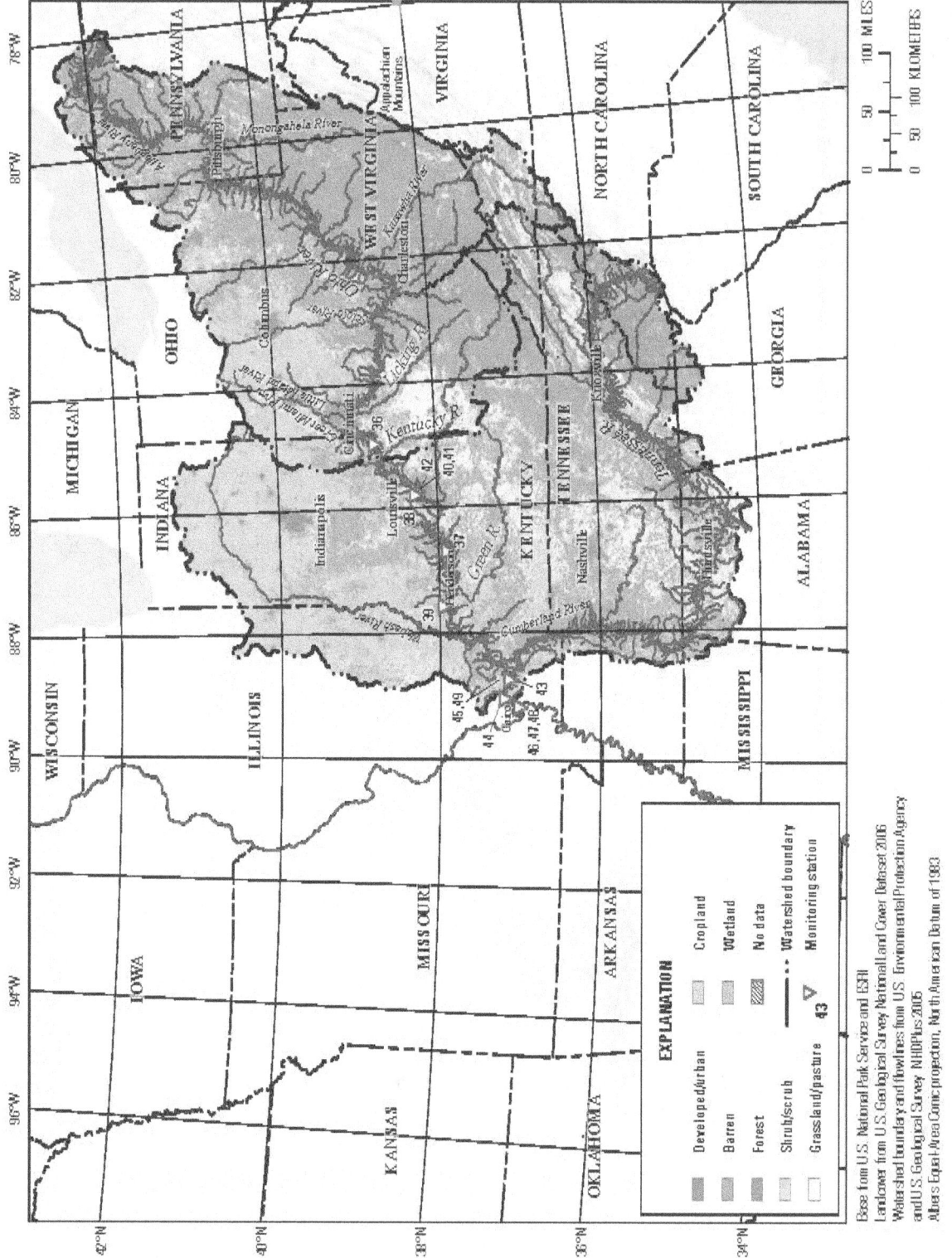

Figure 19. Ohio River basin, Ohio.

This achievement occurred at a time when water quality was deteriorating dramatically in many other parts of the country. At the time of formation, ORSANCO relied heavily upon a massive survey of pollution in the Ohio River basin conducted by the U.S. Public Health Service in 1940–42. The published report was nearly 1,400 pages long and detailed water quality, pollution sources, recommendations for remediation, and cost estimates for the entire basin (U.S. Public Health Service, 1942). The principal pollution issues identified by the Ohio River Pollution Survey are familiar, but instructive. The report included recommended standards for coliform bacteria, dissolved oxygen, biochemical oxygen demand, pH, and phenol content (Cleary, 1967). In the 1940s, the main driving forces of pollution in this basin were identified as deriving from industry, population centers, and mining.

In contrast to many heavily farmed basins included in this study, the Ohio River basin is approximately 47 percent forested land and 47 percent agricultural land (Benke and Cushing, 2005). Meanwhile population density for the entire basin (referenced as the lower Ohio River) is 53 people/km², and comparable for the middle basin upstream of Louisville at 62 people/km² (fig. 3). These values reflect a mix of large population centers with more sparsely populated agricultural and forest land. Agricultural intensity differs between the upper and lower part of the Ohio River basin with the percentage of the basin in agricultural land usage ranging from less than 20 percent in Pennsylvania in to more than 70 percent in Illinois and Indiana, near the confluence with the Mississippi (Benke and Cushing, 2005). Slightly more than 45 percent of the entire Ohio River basin is agricultural land with 30 percent in cropland (fig. 4). Upstream of Louisville, Ky., the basin is 37 percent farmland with 22 percent in cropland (fig. 4). The basin has steadily lost farmland throughout the 20th century with nearly 80 percent of the basin in farmland and 55 percent of the basin in cropland in 1900 (figs. 16C–16D). Industrial and urban sources are likely to be the principal driving forces of pollution upstream of Louisville, Ky., while agricultural influences may be more significant in the lower part of the Ohio River basin.

Data Sources

For the purposes of our study, two reaches of the Ohio River were considered. The middle Ohio River was assessed at stations around Louisville, Ky. and Evansville, Ind. The lower Ohio River includes data from stations around Cairo and Metropolis, Ill. Data sources for the lower and middle Ohio River are summarized in table 11.

For the middle Ohio River, daily streamflow is available from 1928 to 2010 at the reference stream gaging station at Louisville, Ky. (station No. 03294500) and from the 1970s to 2007 from Cannelton Dam. The drainage area at Louisville, Ky., is about 236,000 km² (91,000 mi²), mean annual discharge for the period 1970–2007 is 3,300 m³/s (116,000 ft³/s), and runoff is about 440 mm/yr. In 1968, USGS water-quality sampling began at the reference stream gaging station at Louisville. Water-quality data also are available from the USGS stream-gaging station farther downstream at Cannelton Dam and from the STORET database from Evansville. Predating the USGS data on the middle Ohio River are data collected by the U.S. Public Health Service in 1939–40 as part of the Ohio River Pollution Survey and in 1929–30 (monitoring station 42) as part of a follow-up study to the groundbreaking work done in 1914–15 on the Ohio River. The earliest water–quality data on the Middle Ohio River comes from State and local sources. The Indiana State Board of Health conducted a longitudinal survey on the Ohio River in 1911 that contains water–quality data for a number of stations. The Louisville Water Company also published water-quality data collected from the Ohio River at Louisville, Ky., in 1895–1897.

For the lower Ohio River, daily streamflow data began to be collected at the reference stream gaging station at Metropolis, Ill. (station No. 03611500), in 1928. The USGS began collecting water-quality data at Dam 53 near Grand Chain, Ill., in the mid-1950s. The Illinois State Water Survey also collected water-quality data at Cairo, Ill., from 1958 to 1976 and at Metropolis, Ill., from 1950 to 1956. These data were published in the Illinois State Water Survey Bulletin. The oldest data on the lower Ohio River also were collected by the Illinois State Water Survey as part of the work begun by Arthur Palmer in the late 19th century. Data collected at Cairo and Metropolis, Ill., exist for a few years between 1898 and 1908.

Table 11. Ohio River basin data sources.

[**Monitoring station** is shown in figure 19. **Source:** NWIS, National Water Information System; STORET, Storage and Retrieval Data Warehouse. **Constituent:** Alk, alkalinity. See table 2 for definitions of all other constituent abbreviations]

Monitoring station	Source	Station	Latitude	Longitude	Constituent	Start year	End year
			Middle Ohio River				
36	NWIS	03277200 Ohio River at Markland Dam, near Warsaw, Ky.	38°46′29″	84°57′52″	Daily streamflow	1970	2007
					TN	1974	1986
					NH_3	1977	1986
					$TON + NH_3$	1974	1986
					NO_3	1959	1986
					Alk	1959	1986
					TP	1967	1986
					TDP	1978	1986
37	NWIS	03303280 Ohio River at Cannelton Dam, at Cannelton, Ind.	37°53′58″	86°42′20″	Daily streamflow	1975	2007
					TN	1974	2009
					NH_3	1977	2009
					$TON + NH_3$	1974	2009
					NO_3	1974	2009
					Alk	1974	2009
					TP	1974	2009
					TDP	1978	2009
[1]38	NWIS	03294500 Ohio River at Louisville, Ky.	38°16′49″	85°47′57″	Daily streamflow	1928	2010
					NO_3	1968	1995
					Alk	1968	1995
					TP	1968	1995
39	STORET	170036 Ohio River at Evansville, Ind.	37°57′28″	87° 4′28″	TN	1960	1974
					NH_3	1958	1974
					$TON + NH_3$	1960	1974
					NO_3	1960	1974
					Alk	1958	1974
					TP	1960	1974
					TDP	1964	1974
[2]40	Fuller, 1898	Ohio River at Louisville, Ky.	38°16′49″	85°47′57″	NH_3	1895	1897
					NO_3	1895	1897
					Alk	1895	1897
[2]41	State Board of Health of Indiana, 1911	Ohio River at Louisville, Ky.	38°16′ 49″	85°47′57″	NH_3	1911	1911
					NO_3	1911	1911
					Alk	1911	1911
[2]42	Crohurst, 1933	Ohio River at Louisville, Ky.	38°15′ 49″	85°45′06″	Alk	1929	1930

Table 11. Ohio River basin data sources.—Continued

[**Monitoring station** is shown in figure 19. **Source:** NWIS, National Water Information System; STORET, Storage and Retrieval Data Warehouse. **Constituent:** Alk, alkalinity. See table 2 for definitions of all other constituent abbreviations]

Monitoring station	Source	Station	Latitude	Longitude	Constituent	Start year	End year
			Lower Ohio River				
[1]43	NWIS	03611500 Ohio River at Metropolis, Ill.	37°08'51"	88°44'27"	Daily streamflow	1928	2008
					TN	1978	1995
					NH_3	1978	1995
					$TON + NH_3$	1978	1995
					NO_3	1978	1995
					Alk	1978	1995
					TP	1978	1995
					TDP	1978	1995
44	NWIS	03612500 Ohio River at Dam 53, near Grand Chain, Ill.	37°12'11"	89°02'30"	TN	1972	2009
					NH3	1972	2009
					$TON + NH_3$	1972	2009
					NO_3	1954	2009
					Alk	1954	2009
					TP	1967	2009
					TDP	1972	2009
[2]45	Larson and Larson, 1957	Ohio River at Metropolis, Ill.	37°08'51"	88°44'27"	NH_3	1950	1956
					NO_3	1950	1956
					Alk	1950	1956
[2]46	Larson and Larson, 1957; Harmeson and Larson, 1969; Harmeson and others, 1973	Ohio River at Cairo, Ill.	37°12'11"	89°10'30"	NH_3	1958	1976
					NO_3	1958	1976
					Alk	1958	1976
					TDP	1960	1976
[2]47	Long, 1889	Ohio River at Cairo, Ill.	37°01'23"	89°10'30"	NO_3	1888	1888
					NH_3	1888	1888
[2]48	Bartow, 1907, 1909	Ohio River at Cairo, Ill.	37°01'23"	89°10'30"	NH_3	1898	1908
					NO_3	1898	1908
					Alk	1898	1908
[2]49	Bartow, 1907, 1909	Ohio River at Metropolis, Ill.	37°08'41"	88°44'32"	NH_3	1898	1908
					NO_3	1898	1908
					Alk	1898	1908

[1]Reference stream gaging station.

[2]Historical station, location approximate.

Illinois River

The Illinois River is a major tributary to the upper Mississippi River with the confluence of the two rivers located at Grafton, Ill. (fig. 20). The Illinois River basin covers more than 75,000 km² (29,000 mi²) and lies almost entirely within Illinois with small portions in Indiana and Wisconsin.

The main stem is approximately 440 km long and contributes nearly 20 percent of the total water discharge in the upper Mississippi at Alton, Ill. The farthest downstream gaging station is at Valley City, Ill., almost 100 km upstream of the confluence because river stage downstream from there can be affected by flooding on the Mississippi River (Groschen and others, 2000). Mean annual discharge is approximately 650 m³/s (23,000 ft³/s) and runoff is 300 mm/yr at the reference stream gaging station for the lower basin at Valley City (station No. 05560000) for the period 1938–2010. The river is formed by the convergence of the Des Plaines and Kankakee Rivers southwest of Chicago, Ill. Other tributaries include the Fox, Vermillion, Mackinaw, Spoon, Sangamon, and La Moine Rivers. The Illinois River basin lies almost entirely within the Till Plains section of the Central Lowlands physiographic province. The lower Illinois basin is heavily cultivated with farmland making up more than 80 percent of the land area (fig. 4). In addition, five urban areas with populations greater than 100,000 are within the basin and population density of the lower Illinois basin is 155 people/km² (fig. 3). Both of these metrics indicate very intense human activity, including agricultural production and urbanization in this drainage basin. Therefore, both urban and agricultural pressures are expected to influence water quality in this basin.

The environmental history of the Illinois River is exceptionally well-documented and encompasses several serious environmental challenges faced by this ecosystem. A survey of water pollution on the Illinois River was conducted in 1888 and published in 1889 by the Illinois State Board of Health (Long, 1889) and contains the oldest water chemistry data available for this study. Biological surveys of the aquatic ecosystems of Illinois began in 1894 by Stephen Forbes of the Illinois Natural History Survey (Benke and Cushing, 2005). Around this time, the Illinois State Water Survey also began methodically categorizing the quality of drinking-water sources for municipalities throughout the State under the direction of Arthur Palmer. The first "Chemical Survey of the Water Supplies of Illinois" (Palmer, 1897) was published in 1897 and included water-quality data for groundwater and surface waters throughout the State. These sources contribute greatly to our knowledge of the Illinois River in the late 19th and early 20th centuries.

Perhaps the most significant environmental alteration to the river was the reversal of the Chicago and Calumet Rivers in the late 19th century. These rivers formerly flowed into Lake Michigan and were used extensively at this time as sanitary sewers to transport wastewater away from Chicago. However, Lake Michigan also was the drinking water source for Chicago, so legitimate concerns were raised about the wisdom of dumping wastewater in close proximity to the location of city drinking water intakes. City managers decided to reverse the flow of the Chicago River to direct wastewater into the Illinois River rather than Lake Michigan. Part of the plan also included diluting the wastewater with water pumped from Lake Michigan as a way of ameliorating some of the negative effects of diverting sewage into the Illinois River. The Chicago Sanitary and Shipping Canal opened in 1900 and is still in operation. The alteration added around 2,000 km² (770 mi²) to the river basin and in some years more than 25 percent of the total flow in the Illinois River originated from the canal as either wastewater or water from Lake Michigan meant to dilute the waste (Murphy, 1961).

Although the Illinois River was hardly in pristine condition at the opening of the Chicago Sanitary and Ship Canal, the influx of polluted wastewater accelerated water-quality degradation in the river. Other pressures on the river ecosystem included overfishing, urban and industrial development within the basin, and the development of levee and drainage districts, which disconnected the river from its riparian ecosystem and allowed its extensive bottomland floodplain to be converted into agricultural land (Thompson, 2002). Additionally, the Illinois River once supported diverse and extremely productive fisheries. The low hydrologic gradient and wide floodplain in the lower basin allow long water residence time in the river basin and probably contributed to the magnitude and diversity of its fisheries production under pristine conditions (Benke and Cushing, 2005). Overfishing, low oxygen resulting from pollution inputs, and silt deposition all contributed to the demise of the fisheries on the Illinois River. The mussel fishery collapsed in the early 20th century and the fin fishery collapsed by the 1920s (Thompson, 2002). By 1922 "animal life [had been] almost excluded from the upper river...due to increase of sewage incident to the growth of the city of Chicago" (Purdy, 1930, p. 9).

A great deal of effort has gone into cleaning up the Illinois River, but pressures on the ecosystem are still tremendous. Currently (2012), 196 wastewater-treatment plants exist in the upper drainage basin (Benke and Cushing, 2005) although the influx of municipal wastes has decreased substantially. Nonetheless, nutrient (N and P) concentrations within the river are still very high (Groschen and others, 2000). The lower Illinois River had the highest corn harvest, more than 16,000 bushels/km² of drainage area (fig. 6), and among the highest identified sources of nitrogen and phosphorus of the basins included in this study (figs. 5A–5B).

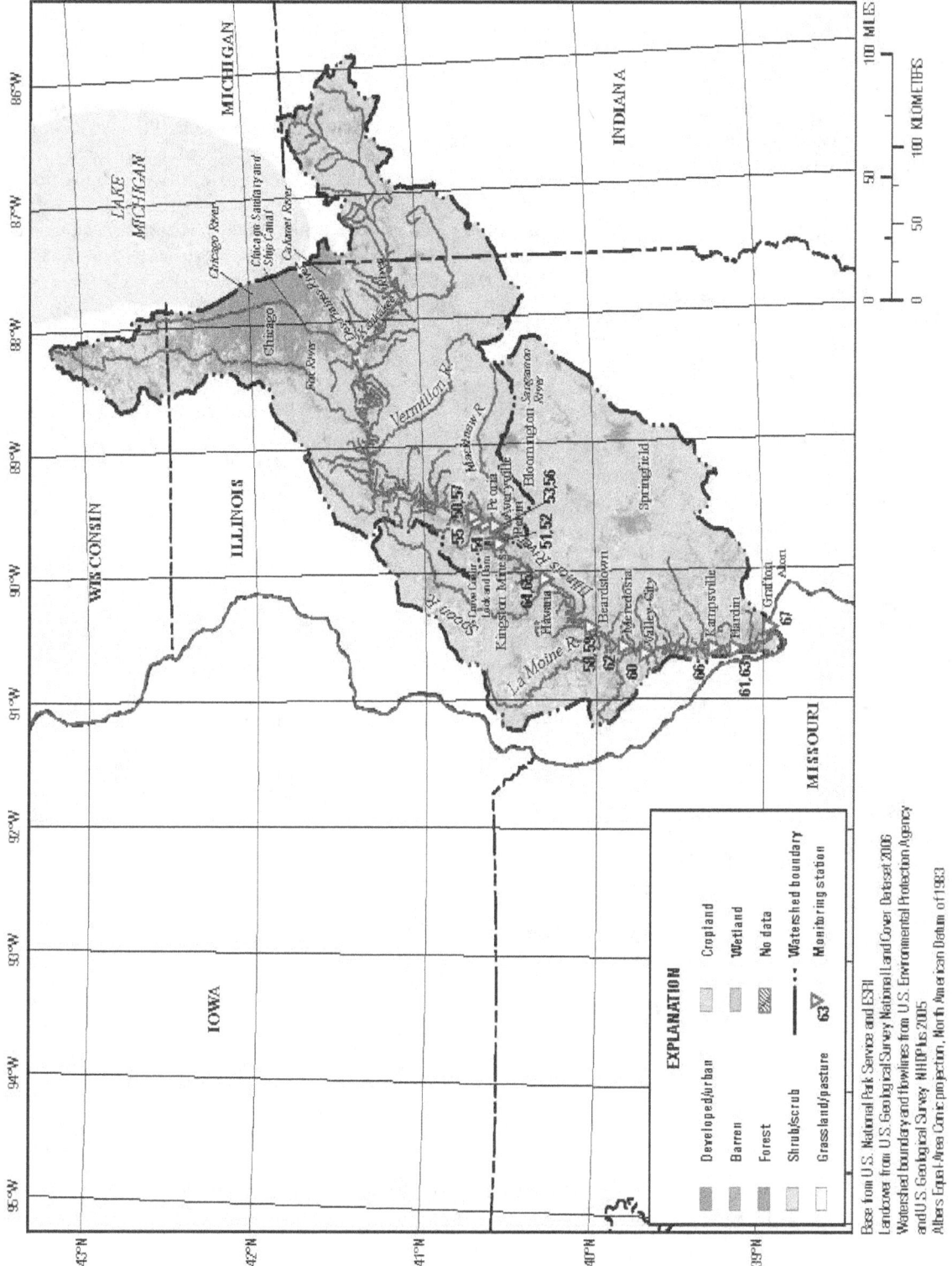

Figure 20. Illinois River basin.

Data Sources

For the purposes of this study, water quality and streamflow data will be considered from two reaches of the Illinois River based on the location of USGS stream gages. The middle Illinois River is defined by streamflow data from around Kingston Mines and the reference stream gaging station at Peoria, Ill. (station No. 05560000), while the lower Illinois River is described by streamflow data from around Beardstown and the reference stream gaging station at Valley City, Ill (station No. 05586100). Data sources for the lower and middle Illinois River are summarized in table 12.

Partial daily streamflow data are available for the middle Illinois River at Peoria beginning in 1903 and complete data are available beginning in 1910 through 1938. Beginning in 1939, the stream-gaging station was relocated approximately 15 mi downstream to Kingston Mines and is still in operation. The drainage area at Kingston Mines incorporates the Mackinaw River drainage basin and is about 41,000 km² (15,800 mi²), about 10 percent larger than Peoria at 36,500 km² (13,700 mi²). For the lower Illinois River, daily streamflow data exist beginning in 1920 at Beardstown, Ill. Since 1938, daily streamflow data have been collected about 30 mi downstream at Valley City, Ill. The drainage area at Valley City is 69,264 km² (26,740 mi²), about 9 percent larger than Beardstown at 62,753 km² (24,230 mi²).

As mentioned above, water-quality data exist as far back as 1888 for both the middle and lower Illinois River. Data also exists for these river reaches as part of volumes 1 and 2 of the Illinois State Water Survey Bulletin around Kampsville in the lower reach and Peoria in the middle reach. Notably, these samples contain Kjeldahl N data, which is still in use today and provides a basis for comparing long-term changes in total N in the Illinois River. Data from 1906 to 1907 at Peoria and Kampsville, Ill. also were included in the Clarke (1924) compilation. A survey of the Illinois River was conducted in 1921–22 and included 23 stations between Lockport (in the Chicago area) and Kampsville, Ill. This study also includes Kjeldahl N measurements and provides critical insight into the total N status of the river before substantial wastewater treatment had begun. For the purposes of this compilation, data from stations at Peoria and Averyville, Ill., were identified as describing conditions for the middle reach, and three stations between Hardin and Beardstown, Ill. (combined and designated as monitoring station 63; table 12) for the lower reach. The Illinois State Water Survey resumed collecting and publishing water-quality data from surface and groundwater in 1945, including the Illinois River at Peoria, which was sampled regularly by the Illinois State Water Survey until 1971. Beginning in 1955, the Illinois River also was sampled at Meredosia, which is between Beardstown and Valley City, Ill. in the lower Illinois River. The Illinois Environmental Protection Agency began collecting data throughout the State in 1958 and continuing through 1977. These data were published by the U.S. Geological Survey and include data from both the lower and middle Illinois River (Healy and Toler, 1978; Grason and Healy, 1979). The USGS began collecting water-quality data at Pekin, Ill., in 1977 and at Valley City, Ill., in 1974. Sampling continued at Valley City, Ill. (2008), but sampling at Pekin was discontinued in 1998, although data from the Illinois EPA Ambient Water Quality Monitoring Network were collected at Pekin, Ill., between 1999 and 2005.

The sampling station at Peoria, Ill., is located near the center of the city and upstream of the major sewer outfall. Therefore, data collected more recently from Pekin, Ill., may include some effect of this point source. However, data are available from the sampling station at Peoria, Ill., beginning in the early 20th century through 1977, so this station offers more than 70 years of data collected from the same site thus allowing for interpretation of long-term trends.

Table 12.   Illinois River basin data sources.

[**Monitoring station** is shown in figure 20. **Source:** NWIS, National Water Information System; STORET, Storage and Retrieval Data Warehouse. **Constituent:** Alk, alkalinity. See table 2 for definitions of all other constituent abbreviations]

Monitoring station	Source	Station	Latitude	Longitude	Constituent	Start year	End year
			Middle Illinois River				
[1]50	NWIS; Clarke, 1924	05560000 Illinois River at Peoria, Ill.	40°42'08"	89°33'52"	Daily streamflow	1903	1938
					NO_3	1906	1907
					Alk	1906	1907
51	NWIS	05568500 Illinois River at Kingston Mines, Ill.	40°33'11"	89°46'38"	Daily streamflow	1939	2010
[2]52	Long, 1889	Illinois River at Pekin, Ill.	40°42'08"	89°33'52"	NO_3	1888	1888
					NH_3	1888	1888
53	NWIS	05563800 Illinois River at Pekin, Ill.	40°34'23"	89°38'17"	TN	1977	1998
					NH_3	1977	1998
					$TON + NH_3$	1977	1998
					NO_3	1977	1998
					Alk	1978	1998
					TP	1980	1998
					TDP	1981	1998
54	STORET	48367 Illinois River at Creve Coeur Lock and Dam, Ill.	40°37'57"	89°37'28"	TN	1989	1990
					NH_3	1974	1990
					$TON + NH_3$	1989	1990
					NO_3	1972	1990
					Alk	1989	1990
					TP	1972	1990
					TDP	1989	1990
55	Larson and Larson, 1957; Harmeson and Larson, 1969; Harmeson and others, 1973	48809 Illinois River at Peoria, Ill.	40°40'55"	80°36'04"	NH_3	1945	1971
					NO_3	1945	1971
					Alk	1945	1971
					TDP	1959	1971
[2]56	Hoskins and others, 1927	Illinois River at Pekin, Ill.	40°34'23"	89°38'17"	TN	1921	1922
					NH_3	1921	1922
					$TON + NH_3$	1921	1922
					NO_3	1921	1922
[2]57	Palmer, 1897, 1903	Illinois River at Peoria, Ill.	40°41'57"	89°34'11"	TN	1896	1901
					NH_3	1895	1901
					$TON + NH_3$	1896	1901
					NO_3	1895	1901

Table 12. Illinois River basin data sources.—Continued

[**Monitoring station** is shown in figure 20. **Source:** NWIS, National Water Information System; STORET, Storage and Retrieval Data Warehouse. **Constituent:** Alk, alkalinity. See table 2 for definitions of all other constituent abbreviations]

Monitoring station	Source	Station	Latitude	Longitude	Constituent	Start year	End year
			Lower Illinois River				
[2]58	Long, 1889	Illinois River at Beardstown, Ill.	40°01′23″	90°26′12″	NO$_3$	1888	1889
					NH$_3$	1888	1889
59	NWIS	05584000 Illinois River at Beardstown, Ill.	40°01′23″	90°26′12″	Daily streamflow	1920	1938
[1]60	NWIS	05586100 Illinois River at Valley City, Ill.	39°42′12″	90°38′43″	Daily Streamflow	1938	2010
					TN	1974	2008
					NH$_3$	1975	2008
					TON + NH$_3$	1974	2008
					NO$_3$	1974	2008
					Alk	1974	2008
					TP	1974	2008
					TDP	1977	2008
61	NWIS	05587060 Illinois River at Hardin, Ill.	39°09′37″	90°36′55″	TN	1978	1995
					NH$_3$	1977	1998
					TON + NH$_3$	1978	1995
					NO$_3$	1977	1998
					Alk	1977	1995
					TP	1978	1998
					TDP	1984	1998
62	Harmeson and Larson, 1969; Harmeson and others, 1973	Illinois River at Meredosia, Ill.	39°49′24″	90°34′05″	NH$_3$	1955	1971
					NO$_3$	1955	1971
					Alk	1955	1971
[2]63	Hoskins, 1927	Illinois River at Hardin, Ill.	39°09′37″	90°36′55″	TN	1921	1922
					NH$_3$	1921	1922
					TON + NH$_3$	1921	1922
					NO$_3$	1921	1922
[2]64	Palmer, 1897, 1903	Illinois River at Havana, Ill.	40°17′40″	90°04′12″	TN	1896	1900
					NH$_3$	1895	1900
					TON + NH$_3$	1896	1900
					NO$_3$	1895	1900
[2]65	Clarke, 1924	Illinois River at Kampsville, Ill.	39°17′59″	90°36′22″	NO$_3$	1906	1907
					Alk	1906	1907
[2]66	Palmer, 1897, 1903	Illinois River at Kampsville, Ill.	39°17′59″	90°36′22″	TN	1896	1901
					NH$_3$	1896	1902
					TON + NH$_3$	1896	1901
					NO$_3$	1896	1902
[2]67	Palmer, 1903	Illinois River at Grafton, Ill.	38°58′04″	90°26′04″	TN	1899	1901
					NH$_3$	1899	1902
					TON + NH$_3$	1899	1901
					NO$_3$	1899	1902

[1]Reference stream gaging station.

[2]Historical station, location approximate.

Des Moines River

The Des Moines River is a tributary to the upper Mississippi River located primarily in Iowa. It is the largest river basin in Iowa, covering nearly 37,500 km² (14,500 mi²). The headwaters are located in southern Minnesota and the farthest southern reaches of the basin are in Missouri (fig. 21). The Des Moines River runs primarily from northwest to southeast and enters the upper Mississippi River just downstream of Keokuk, Iowa. Measured at the reference stream gaging station at Keosauqua, Iowa (USGS station No. 05490500), mean annual discharge on the Des Moines River was approximately 257 m³/s (9,100 ft³/s) between 1970 and 2010 but was extremely variable, ranging from 37 m³/s (1,300 ft³/s) in 1977 to 763 m³/s (almost 27,000 ft³/s) in 1993. Mean annual runoff during this period was 220 mm/yr, but annual runoff also ranged widely from 32 to 660 mm/yr.

The entire basin lies within the Central Lowlands physiographic province and is heavily agricultural, but there are important differences between the northern and southern portions of the basin, basically upstream and downstream of Des Moines, Iowa. The northern part of the basin is located within the Des Moines Lobe landform region and was glaciated 12,000–15,000 years ago. The topography in this region is flat with poorly developed soil drainage networks. Wetlands and prairie potholes covered this area before European settlement and the development of agriculture depended upon building artificial drainages such as ditches and tiles. The relatively young soil is erodible resulting in significant down cutting along the main stem Des Moines River, which causes the channel to be relatively narrow with a constricted alluvial plain. The soils in this area also are highly productive and cultivated for crops, such as corn and soybeans. Downstream of Des Moines, the river flows through the Southern Iowa Drift Plain, which is characterized by older surface soils and highly developed surface drainage networks, including numerous stream channels and steep hill slopes (Keeney and DeLuca, 1993). The main channel of the Des Moines River runs through a wide alluvial plain. Outside of the alluvial plain, older upland soils are not as productive as in the northern part of the basin and so are not as heavily cultivated (Heusinkveld, 1989). The wide, flat alluvial plain in this area also is subject to heavy flooding, which was a constant danger to communities located near the Des Moines River until the construction of upstream flood-control structures (Heusinkveld, 1989).

Many of the oldest communities in southern Iowa are located adjacent to the Des Moines River because it was a critical means of transportation during settlement of this area. However, many of these communities experienced periodic and sometimes devastating flooding. In the latter half of the 20th century, the Saylorville and Red Rock Dams were constructed to help control flooding on the Des Moines River. The Red Rock Dam is located downstream of Des Moines. Construction began in 1960 and was completed in 1969. The Saylorville Dam is upstream of Des Moines. Construction began in 1965 and the reservoir was filled by 1977 (Heusinkveld, 1989). The presence of the Red Rock and Saylorville Dams has helped to curtail the floods that periodically struck communities adjacent to the river.

Among the river basins considered in this analysis, the Des Moines River basin is one of the most intensively cultivated and the population density is relatively low. In 2002, almost 90 percent of the entire basin was in some form of agricultural usage and almost 77 percent was in cropland (fig. 4). Corn and soybeans are the major crops grown in this basin (Keeney and DeLuca, 1993; Benke and Cushing, 2005) and the annual harvest of corn averaged nearly 15,000 bushels/km² of drainage area between 2000 and 2010 (fig. 6). The average annual nitrogen sources from fertilizer and animal livestock between 1997 and 2001 were 8.1 g (N/m²)/yr of drainage area, about 75 percent of which was from fertilizer application (fig. 5A). Phosphorus sources over the same period were nearly 1.6 g (P/m²)/yr and nearly evenly split between animal livestock and fertilizer sources (fig. 5B). In the year 2000 Population Census (U.S. Census Bureau, 2000), human population density in this drainage basin was 21 people/km², less than the national average of 31 (fig. 3). The City of Des Moines, Iowa, is the only large urban area in the basin.

Data Sources

Data sources for the Des Moines River are summarized in table 13. The USGS began stream gaging on the Des Moines River in 1903 at the reference stream gaging station at Keosauqua (station No. 05490500) and at Ottumwa in 1917 (station No. 05489500). The earliest water-quality samples come from Keosauqua in 1906 and were published in the Clarke (1924) national water-quality compilation. Water-quality data also were collected at Keosauqua in 1967 and then resumed in 2004. The Iowa Department of Public Health Engineering published water-quality data collected along the Des Moines River from 1928 to 1934 and the Iowa Department of Natural Resources has data from 16 stations for several years between 1935 and 1955. These same stations were sampled again from 1968 to 1979. The USGS collected water-quality samples regularly at Ottumwa beginning in 1949, continued through 1968, and then intermittently since then. The Des Moines River at St. Francisville, Mo. (station No. 05490600) is the station farthest downstream and was sampled regularly between 1967 and 1993. Because of the shape of the Des Moines River basin, very little additional basin area is gained per river mile in the southernmost part of the basin (fig. 21). For example, even though Ottumwa, Iowa, and St. Francisville, Mo. are relatively far apart in terms of river miles, only 7 percent of the river basin is gained when moving from Ottumwa downstream to St. Francisville, Mo. Therefore, minimal differences in water quality are expected between Ottumwa and St. Francisville.

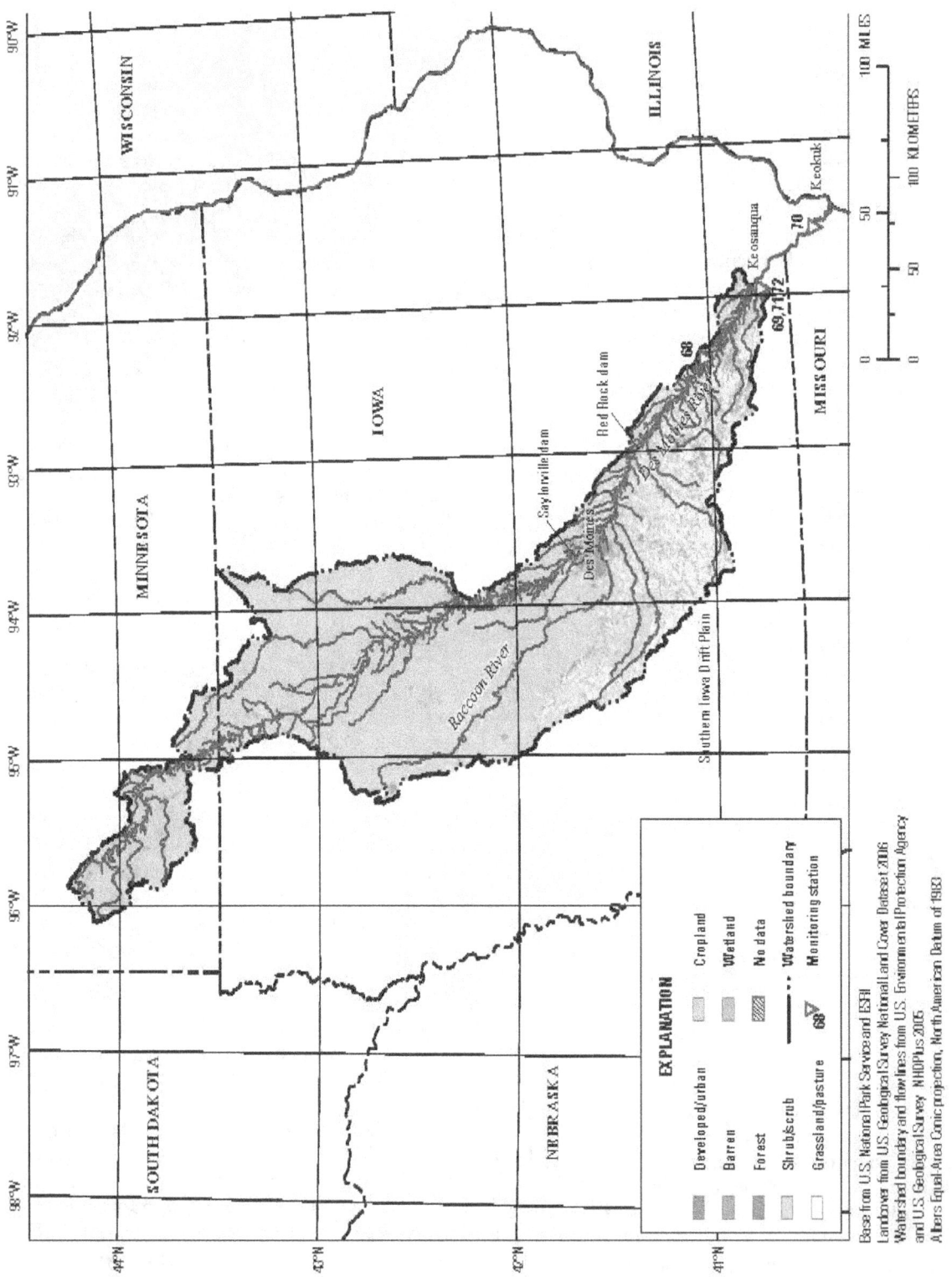

Base from U.S. National Park Service and ESRI
Landcover from U.S. Geological Survey National Land Cover Dataset 2006
Watershed boundary and flowlines from U.S. Environmental Protection Agency
and U.S. Geological Survey NHDPlus 2005
Albers Equal Area Conic projection, North American Datum of 1983

Figure 21. Des Moines River basin.

Table 13. Des Moines River data sources.

[**Monitoring station** is shown in figure 21. **Source:** NWIS, National Water Information System; STORET, Storage and Retrieval Data Warehouse. **Constituent:** Alk, alkalinity. See table 2 for definitions of all other constituent abbreviations]

Monitoring station	Source	Station	Latitude	Longitude	Constituent	Start year	End year
68	NWIS	05489500 Des Moines River at Ottumwa, Iowa	41°0'39"	92°24'41"	Daily streamflow	1917	2008
					TN	2008	2008
					NH_3	1955	2008
					$TON + NH_3$	2008	2008
					NO_3	1949	2008
					Alk	1940	2008
					TP	2008	2008
					TDP	2008	2008
[1]69	NWIS	05490500 Des Moines River at Keosauqua, Iowa	40°43'40"	91°57'35"	Daily streamflow	1903	2008
					TN	2004	2008
					NH_3	1967	2008
					$TON + NH_3$	2004	2008
					NO_3	1906	2008
					Alk	1906	2008
					TP	2004	2008
					TDP	1967	2008
70	NWIS	05490600 Des Moines River at St. Francisville, Mo.	40°27'45"	91°03'41"	TN	1967	1993
					NH_3	1968	1993
					$TON + NH_3$	1967	1993
					NO_3	1967	1993
					Alk	1967	1993
					TP	1967	1993
					TDP	1968	1993
[2,3]71	STORET	410360 Des Moines River at Keosauqua, Iowa	40°43'40"	91°57'35"	TN	1934	[2]1979
					NH_3	1934	[2]1979
					$TON + NH_3$	1934	[2]1979
					NO_3	1934	[2]1979
					Alk	1969	[2]1979
					TP	1968	[2]1979
[2]72	Iowa Division of Public Health Engineering, 1934	Des Moines River at Keosauqua, Iowa	40°43'42"	91°57'39"	NH_3	1928	1934
					NO_3	1928	1934

[1]Reference stream gaging station.

[2]Historical station, location approximate.

[3]Overall time series length contains many missing years.

Missouri River

The Missouri River is a major tributary to the Mississippi River accounting for approximately 20 percent of the Mississippi River total discharge. The Missouri River drainage basin covers about 1.37 million km² (529,000 mi²) of the north-central Great Plains and the Eastern Slope of the Rocky Mountains (fig. 22). Parts of 10 states, Canada, and 7 physiographic provinces are within the Missouri River drainage basin (fig. 2). Accordingly, climatic conditions vary widely from cold and moist in the Rocky Mountains to semiarid in the Great Plains to humid continental in the Central Lowlands. Mean annual discharge near the confluence with the Mississippi River nearly doubles the mainstem streamflow at that point, measured at the reference stream gaging station at Hermann Mo. (station No. 06934500) for the period 1928–2008 as 80,000 ft³/s. This translates to annual runoff for the entire basin as 52 mm/yr and reflects the prominence of the dry Great Plains located within the basin.

Historically, the Missouri River was shallow and sediment-laden with frequently shifting channels, riparian sloughs, and backwaters (Sprague and others, 2007). Historically the river experienced large interannual variation in discharge because of the contribution of melting snow in the northern Rocky Mountains to overall river flow. Management of the Missouri River, including snag removal and bank stabilization dates back to the 19th century. However, the most intensive management steps began with the initiation of the Pick-Sloan Missouri Basin Project in 1944 and the Missouri River Bank Stabilization and Navigation Project of 1945. These projects were officially completed in 1981 with a large portion of the activity occurring in the 1950s (U.S. Geological Survey, 1998). The projects aimed to provide flood and erosion protection to infrastructure, communities, and agricultural lands in the Missouri River basin as well as a navigable channel along the lower 1,180 km of the river from St. Louis, Missouri to Sioux City, Iowa. Additionally, reservoirs and dams were constructed throughout the basin for flood control and to provide irrigation water. The total storage capacity of this network is more than 9×10^{10} m³ (more than 73 million acre-ft), making it the largest reservoir network in North America (Sprague and others, 2007). Engineering the Missouri River has dramatically altered the annual hydrograph and has greatly reduced the amount of sediment delivered to the Mississippi River (Meade and Moody, 2010). Bank stabilization projects also greatly curtailed the periodic inundation of wetlands and backwaters, which originally promoted ecosystem productivity and supported diverse native fish and migratory bird populations (U.S. Geological Survey, 1998).

Prior to the 1950s, the Missouri River delivered an estimated 300 million tons of sediment to the Mississippi River, approximately 75 percent of the total sediment load carried to coastal Louisiana (Meade and Moody, 2010). However, by 1980, the Missouri River carried less than 100 million tons of sediment and the entire Mississippi River basin delivered approximately 150 million tons of sediment to the Louisiana coast (Meade and Moody, 2010).

Population density throughout the entire basin was very low according to the 2000 census (U.S. Census Bureau, 2000), 8.7 people/km² (fig. 3), but several large metropolitan areas are within the basin including Denver, Colo., Omaha, Nebr., Kansas City and St. Louis, Mo. In 2002, almost 80 percent of the basin was farmland and this was split nearly evenly between cropland and other uses (fig. 4). Agricultural land development occurred late in this basin with less than 40 percent of the basin in farmland usage in 1900, increasing to 70 percent by 1940 (fig. 17C). The most highly cultivated land occurs adjacent to the main river channel or in the eastern parts of the drainage basin in eastern Kansas, eastern South Dakota, western Iowa, and western Missouri. Average nitrogen sources from fertilizer and animal livestock were moderate compared with other basins in this study, 2.8 g (N/m²)/yr on average from 1997 to 2001 (fig. 5A). Phosphorus sources were moderate as well over the same period, 0.5 g (P/m²)/yr with slightly more phosphorus arising from animal livestock than fertilizer (fig. 5B).

Data Sources

Data sources for the Missouri River are summarized in table 14. The USGS maintains reference stream gaging and water-quality station on the Missouri River at river mile 97.9 near Hermann, Mo. (station No. 06934500, table 14). Nutrient concentration samples have been collected from this site regularly since 1967 and daily streamflow has been collected since 1928 (table 14). This station also was included in the Clarke (1924) publication with nitrate and alkalinity values reported in 1906–07. Monthly average NO_3 and alkalinity concentration data also are available from 1930 to 2008 at the Howard Bend Drinking Water Intake Facility located at river mile 37 near Chesterfield, Mo. (table 14). Because these sites are close along the main stem of the Missouri River and because the drainage area at both sites is nearly identical, 1.35 versus 1.37 million km² at the Hermann and Howard Bend sites, respectively, these sites can be considered equivalent in an assessment of water quality on the Missouri River.

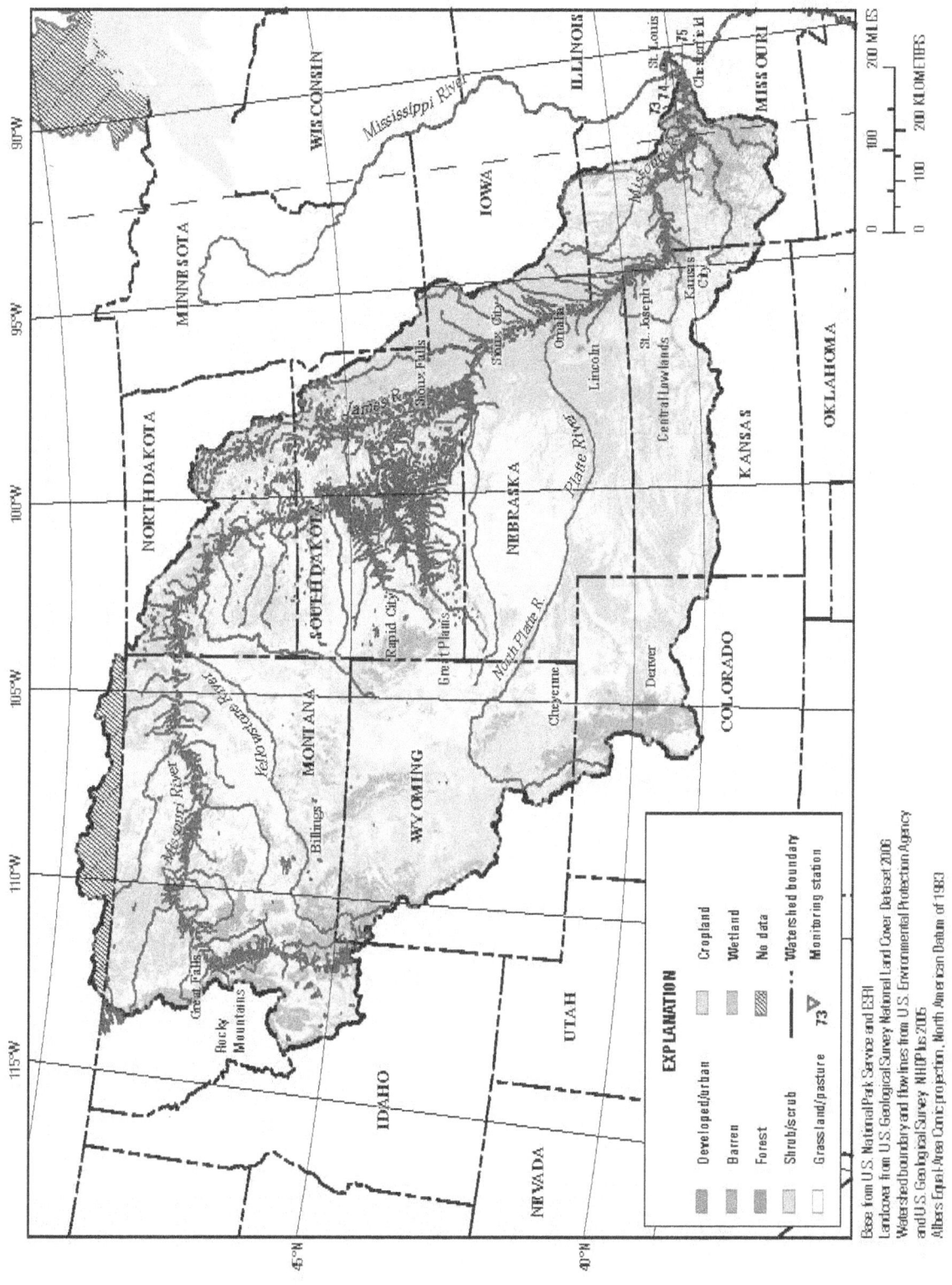

Figure 22. Missouri River basin.

Base from U.S. National Park Service and ESRI
Landcover from U.S. Geological Survey National Land Cover Dataset 2006
Watershed boundary and flow lines from U.S. Environmental Protection Agency
and U.S. Geological Survey NHDPlus 2005
Albers Equal-Area Conic projection, North American Datum of 1983

EXPLANATION

Developed/Urban

Barren

Forest

Shrub/scrub

Grass land/pasture

Cropland

Wetland

No data

Watershed boundary

73 ▽ Monitoring station

Table 14.   Missouri River basin data sources.

[**Monitoring station** is shown in figure 22. **Source:** NWIS, National Water Information System. **Constituent:** Alk, alkalinity. See table 2 for definitions of all other constituent abbreviations]

Monitoring station	Source	Station	Latitude	Longitude	Constituent	Start year	End year
[1]73	NWIS	06934500 Missouri River at Hermann, Mo.	38°42'35"	91°26'19"	Daily streamflow	1928	2010
					TN	1967	2008
					NH_3	1970	2008
					$TON + NH_3$	1967	2008
					NO_3	1967	2008
					Alk	1967	2008
					TP	1967	2008
					TDP	1969	2008
[2]74	Clarke, 1924	06934500 Missouri River at St. Charles (Ruegg), Mo.	38°47'48"	90°27'57"	NO_3	1906	1907
					Alk	1906	1907
75	City of St. Louis Water Division (written commun., Microsoft® Excel file)	Missouri River at Howard Bend Intake, Chesterfield, Mo.	38°40'51"	90°32'40"	NO_3	1930	2008
					Alk	1930	2008

[1]Reference stream gaging station.

[2]Historical station, location approximate.

Arkansas River

The Arkansas River flows primarily east-southeasterly through the south-central plains of the United States and is a major tributary of the Mississippi River (fig. 23). The Arkansas River drainage basin covers more than 154,400 mi² (400,000 km²) and includes parts of six physiographic provinces and seven States (fig. 2). At the reference stream gaging station at Little Rock, Ark. (station No. 07263500), which is near the confluence with the Mississippi River, mean annual discharge (1928–1970) is more than 1,100 m³/s (38,800 ft³/s) and runoff for the entire basin is 87 mm/yr. Major tributaries include the Neosho, Verdigris, Cimarron, and Canadian Rivers. The headwaters of the Arkansas River are located in the southern Rocky Mountains near Leadville, Colo., at approximately 3,000 m (10,000 ft) above sea level. The climate at the headwaters is cold and wet, and annual runoff can reach 90 mm/yr as it emerges from the mountains near Pueblo, Colo. However, evaporation and extensive withdrawals for irrigation reduce the overall flow such that runoff out of the John Martin Reservoir in southeastern Colorado averages only 5 mm/yr and the river becomes ephemeral in western Kansas upstream of Great Bend, Kans. (Benke and Cushing, 2005). At Great Bend, the Pawnee River joins the Arkansas River and increases overall flow. The river gains water through central Kansas and Oklahoma, and after the Cimarron River joins the Arkansas River upstream of Tulsa, Okla., mean annual discharge (USGS station 07164500, 1965–2011) is just under 250 m³/s. Runoff for the entire basin is low because part of the basin is located in arid parts of eastern Colorado, western Kansas, and northern Oklahoma. At the gaging station near Little Rock, approximately 15 percent of the basin (57 km² or 22,000 mi²) is probably non-contributing.

The lower portion of the river has been extensively modified by the locks, dams, and canals making up the McClellan-Kerr Arkansas River Navigation System, which allows commercial traffic on the river and provides recreational opportunities through the creation of reservoirs (Benke and Cushing, 2005). The navigation system begins at the confluence of the White River and the Mississippi River and then joins the Arkansas River near Gillett, Ark., by means of the man-made Arkansas Post Canal. Fourteen lock and dam structures are located along the main stem of the Arkansas River and an additional three extend the system approximately 72 km (45 mi) upstream on the Verdigris River to the Port of Catoosa, Okla. (near Tulsa, Okla.). In all, the system covers more than 400 river miles. Construction began on this project in 1963 and continued into 1970.

Agriculture is a major land use in the river basin with grazing and rangeland uses dominating in arid parts of the western basin; row-cropping becomes more common and more intense in the eastern parts of the drainage basin. In the 1930s, tremendous damage was done to agricultural lands in the basin from the Dust Bowl, including severe wind erosion in eastern Colorado and Kansas as well as moderate to severe sheet and gully erosion in Oklahoma (Hansen and Libecap, 2004). In 2002, even though total farmland was very high in the Arkansas River basin (fig. 4), the intensity of fertilizer use in the basin during a comparable time period was low (figs. 5A–5B) because of the predominance of rangeland and dryland wheat agriculture in the western and central parts of the basin. In the year 2000, population density in the river basin was 15 people/km², about one-half the national average of 31 people/km² (fig. 3).

Data Sources

Data sources for the Arkansas River are summarized in table 15. Long-term data are available around Little Rock, Ark. Daily streamflow data were collected since 1927 at Little Rock (station No. 07263500). After construction of the Murray Dam in 1969 as part of the McClellan-Kerr Arkansas River Navigation System, the gaging station was moved approximately 7 river miles upstream to the outlet of the Murray Dam (considered the reference stream gaging station), renumbered as USGS station No. 07263450, and daily streamflow continued to be collected. Therefore, a good daily streamflow record exists for this river reach dating back to 1927 (table 15). The earliest water-quality samples from this river reach were collected in 1906 and included in the Clarke (1924) compilation of water-quality data collected across the country (table 15). Water-quality sampling by the USGS resumed in 1945 and was collected routinely until 1969 at the Little Rock gaging station (07263500). Beginning in 1969 and continuing through 2009, water-quality data have been collected several river miles downstream at the outlet of the David D. Terry Lock and Dam (07263620). Water-quality data also have been collected from the outlet of Murray Dam. Two sewer outfalls are located in this river reach, which could influence interpretation of water-quality data. The first is upstream of the USGS gaging station at Little Rock and the second is between the Little Rock and David D. Terry gaging stations. The presence of these point sources must be considered when interpreting trends in water quality. Some water-quality data also have been collected around Little Rock, Ark., by the EPA and the Arkansas Pollution Control and Ecology Commission and published in the EPA STORET database (table 15).

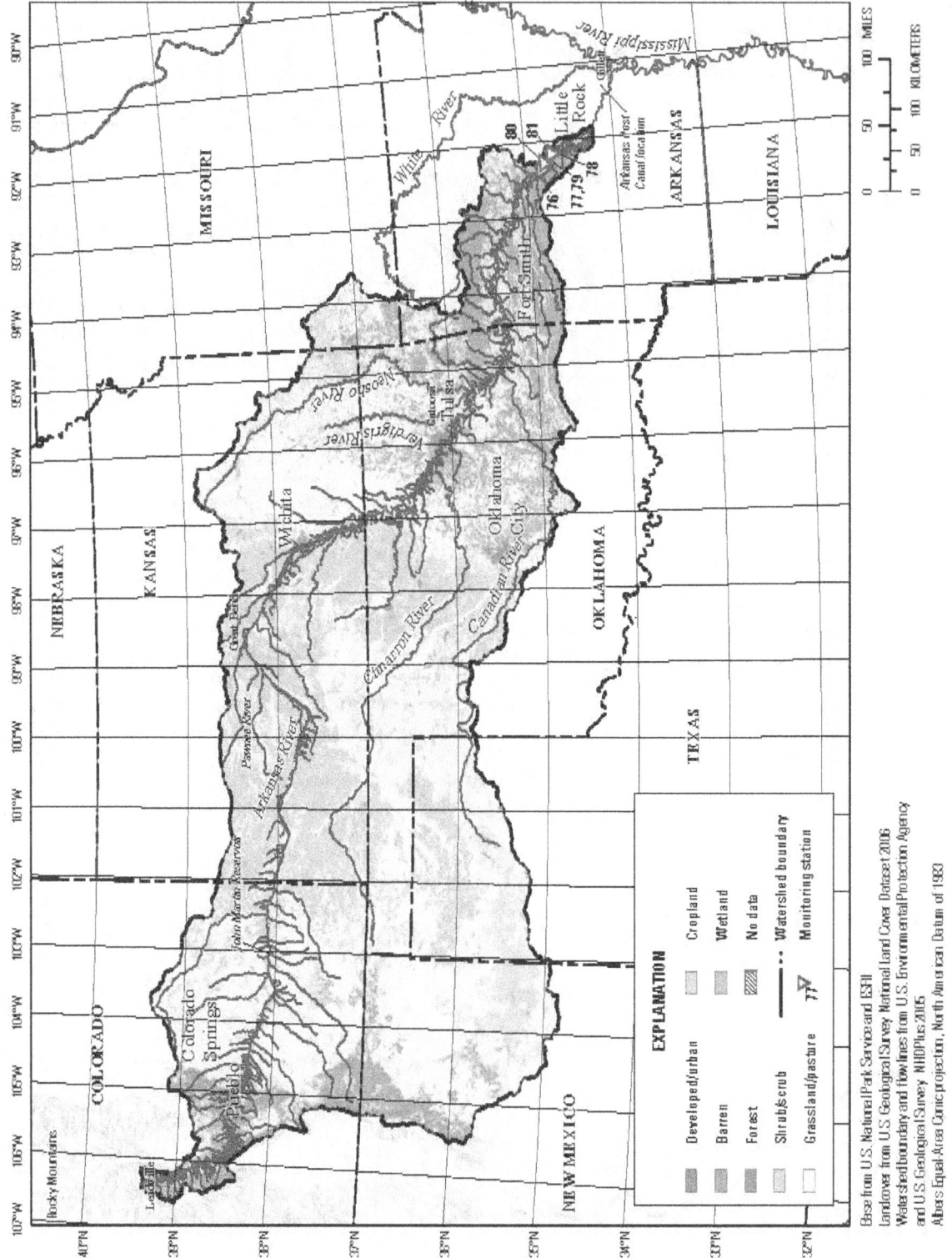

Figure 23. Arkansas River basin.

Table 15. Arkansas River basin data sources.

[**Monitoring station** is shown in figure 23. **Source:** NWIS, National Water Information System; STORET, Storage and Retrieval Data Warehouse. **Constituent:** Alk, alkalinity. See table 2 for definitions of all other constituent abbreviations]

Monitoring station	Source	Station	Latitude	Longitude	Constituent	Start year	End year
76	NWIS	07263450 Arkansas River at Murray Dam, near Little Rock, Ark.	34°47′27″	92°21′33″	Daily streamflow	1970	2010
					TN	1977	1994
					NH_3	1977	1994
					$TON + NH_3$	1977	1994
					NO_3	1975	1994
					Alk	1975	1980
					TP	1975	1994
[1]77	NWIS	07263500 Arkansas River at Little Rock, Ark.	34°44′59″	92°16′09″	Daily streamflow	1927	1970
					TN	1968	1969
					$TON + NH_3$	1967	1969
					NO_3	1945	1969
					Alk	1945	1969
					TP	1965	1969
					TDP	1964	1969
78	NWIS	07263620 Arkansas River at David D. Terry Lock and Dam below Little Rock, Ark.	34°40′52″	92°09′05″	TN	1969	2009
					NH_3	1970	2009
					$TON + NH_3$	1969	2009
					NO_3	1969	2009
					Alk	1969	2009
					TP	1969	2009
					TDP	1977	2009
[2]79	Clarke, 1924	Arkansas River near Little Rock, Ark.	34°44′59″	92°16′09″	NO_3	1906	1907
					Alk	1906	1907
80	STORET	040131 Arkansas River at Little Rock, Ark.	34°46′17″	92°17′49″	TN	1964	1968
					$TON + NH_3$	1964	1968
					NO_3	1964	1968
					Alk	1963	1969
					TP	1964	1969
81	STORET	050056 Arkansas River at David D. Terry Lock and Dam	34°40′07″	92°09′18″	TN	1986	1998
					NH_3	1986	1998
					$TON + NH_3$	1986	1998
					NO_3	1986	1998
					TP	1986	1998

[1]Reference stream gaging station.

[2]Historical station, location approximate.

Brazos River

The Brazos River is a relatively low-gradient stream for most of its length, extending across central Texas from southeastern New Mexico to the Gulf of Mexico (fig. 24). The basin is the second largest in Texas, more than 116,800 km² (45,100 mi²) at the reference stream gaging at Richmond, Tex. (station No. 08114000). Numerous intermittent and relatively saline streams exist in the upper basin and much of this area, approximately 25,000 km² (9,500 mi²), probably is non-contributing to downstream flow in the main stem Brazos River (Vogl and Lopes, 2009). The Brazos River flows through the Great Plains, Central Lowlands, and Coastal Plain physiographic provinces, which are largely characterized by grasslands and prairies (Benke and Cushing, 2005). Precipitation in the basin ranges from about 41 cm (16 in/yr) in the upper basin to more than 127 cm (50 in.) near the river mouth (Wurbs and others, 1993). Mean annual discharge at the reference stream gaging station at Richmond (station No. 08114000) for the period 1903–2010 is approximately 210 m³/s (7,500 ft³/s), which coincides with mean annual runoff of 56 mm/yr from the entire drainage area. Runoff is approximately 5 percent of precipitation in this basin, emphasizing the magnitude of evapotranspiration (Benke and Cushing, 2005). Recurring drought conditions and population growth have been identified as the most important issues for water management in the basin, resulting in the development of a large network of reservoirs located primarily in the middle and upper basin (Vogel and Lopes, 2009).

Rangeland accounted for 42 percent of land cover in 2002 (data not shown), and the primary land-use activities included grazing, cropland, and urban development (fig. 24). The oil and gas industries also are important components of the economy, especially in the upper regions of the basin (Vogl and Lopes, 2009). Agricultural production includes cattle, cotton, grain sorghum, wheat, peanuts, dairy products, and rice (Benke and Cushing, 2005). Forest, pasture, and cropland predominate in the upper basin while grassland and row crops are important farther downstream, sometimes abutting directly on the river's edge with no riparian buffer (Zeng and others, 2011). Nitrogen fertilizer application rates and animal livestock sources of nitrogen both averaged approximately 2 g (N/m²)/yr for 1997–2001 (fig. 5A). However, animal livestock sources of phosphorus were approximately double that of fertilizer sources over the same period (fig. 5B).

Population density in the Brazos River basin was about 18 people/km², based on the 2000 census (fig. 3). Several large cities (population greater than 100,000) are located in the basin (Lubbock, Abilene, and Waco). Urban sprawl in the Houston area also is an important influence near the river mouth, although the center of Houston is located outside the basin. Water use has been dominated by agriculture for most of the century, especially in the middle basin, although population growth since 1970 has been associated with increasing water use for industrial and domestic use (Vogl and Lopes, 2009).

Data Sources

Data sources for the Brazos River are summarized in table 16. Daily streamflow data are available from the reference stream gaging station at Richmond, Tex. (station No. 08114000), beginning in 1903. Water-quality data from 3- to 10-day composite samples are available from selected USGS Water Supply Papers for 1946–58 (Paulsen, 1950; Paulsen, 1952a; Paulsen, 1952b; Paulsen, 1953; U.S. Geological Survey, 1955; Love, 1956; U.S. Geological Survey, 1958; Love, 1959a; Love, 1959b; Love, 1960; Love, 1961; Love, 1963). Water-quality data from point samples are available from NWIS beginning in 1959.

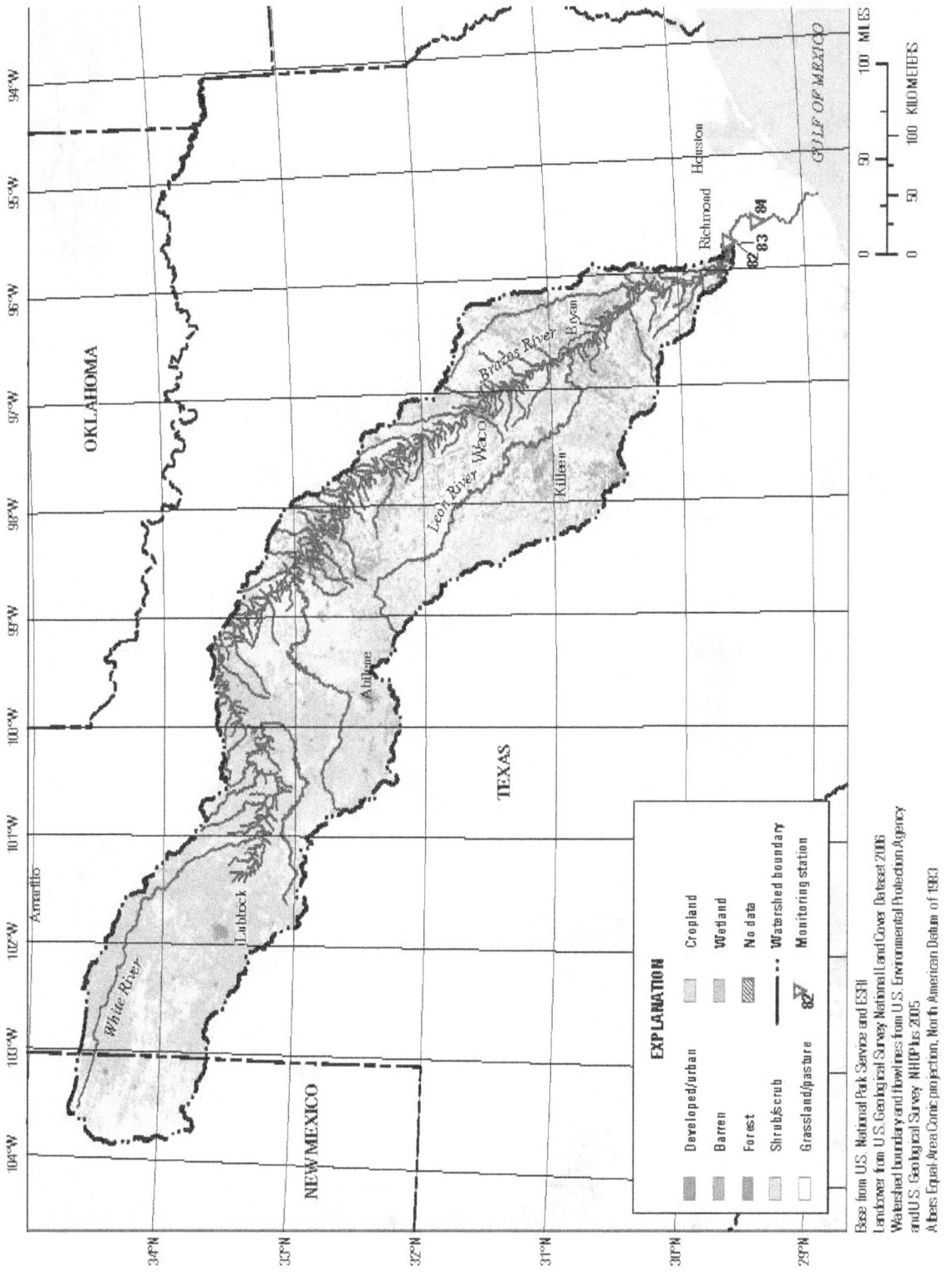

Figure 24. Brazos River basin.

Base from U.S. National Park Service and ESRI
Landcover from U.S. Geological Survey National Land Cover Dataset 2006
Watershed boundary and flowlines from U.S. Environmental Protection Agency
and U.S. Geological Survey NHDPlus 2005
Albers Equal-Area Conic projection, North American Datum of 1983

Table 16.   Brazos River basin data sources.

[**Monitoring station** is shown in figure 24. **Source:** NWIS, National Water Information System. **Constituent**: Alk, alkalinity. See table 2 for definitions of all other constituent abbreviations]

Monitoring station	Source	Station	Latitude	Longitude	Constituent	Start year	End year
[1]82	NWIS	08114000 Brazos River at Richmond, Tex.	29°34'56"	95°45'27"	Daily streamflow	1903	2010
					TN	1971	1995
					NH_3	1970	2002
					$TON + NH_3$	1970	1995
					NO_3	1959	2002
					Alk	1959	2002
					TP	1969	1995
					TDP	1980	2002
[2]83	([3])	08114000 Brazos River at Richmond, Tex.	29°34'56"	95°45'27"	NO_3	1946	1958
					Alk	1946	1958
84	NWIS	08116650 Brazos River at Rosharon, Tex.	29°20'58"	95°34'56"	Daily streamflow	2006	2010
					TN	1970	2008
					NH_3	1970	2008
					$TON + NH_3$	1970	2008
					NO_3	1968	2008
					Alk	1968	2008
					TP	1969	2008
					TDP	1977	2008

[1]Reference stream gaging station.

[2]Historical station, location approximate.

[3]Paulsen, 1950, 1952a, 1952b, 1953a, 1953b, 1953c; U.S. Geological Survey, 1955, 1958; Love, 1959a, 1959b, 1960, 1961, 1963.

Upper Mississippi River

The upper Mississippi River is defined as the part of the Mississippi River upstream of the confluence with the Missouri River near St. Louis, Mo. (fig. 25). The main stem Mississippi River begins at the outlet of Lake Itasca, Minn., and flows approximately 2,000 km to the confluence with the Missouri River in St. Louis, Mo. At Alton, Ill., just upstream of the confluence of the Mississippi and Missouri Rivers, the basin area is 0.44 million km² (170,000 mi²) and incorporates parts of the States of Minnesota, Wisconsin, Iowa, Illinois, and Indiana. Long-term mean annual discharge (1933–1987) at the reference stream gaging station 24 km upstream from this site (station No. 05587450) is approximately 3,100 m³/s (110,000 ft³/s) and runoff is 220 mm/yr.

Most of the basin is located within the Central Lowlands physiographic province, and land use in this basin is characterized by the presence of several large population centers mixed with extensive cultivated agricultural areas (Benke and Cushing, 2005). Large population centers in the upper Mississippi River basin include Chicago, Ill., Minneapolis-St. Paul, Minn., Des Moines, Iowa, and the Quad Cities Metro Area (Bettendorf and Davenport, Ia., and Moline and Rock Island, Ill.) in Iowa and Illinois. Nevertheless, population density throughout the entire basin was only 48 people/km² in 2000, or about 50 percent greater than the national average (fig. 3). This indicates that outside of the major metropolitan areas, population densities were low. Several of the most highly agricultural drainage basins included in this study, the Des Moines River and the Illinois River (figs. 20 and 21), are located within the upper Mississippi River basin. Accordingly, nearly 70 percent of the total area in the upper Mississippi River basin was used for agriculture by the end of the 20th century and the primary use of this land was cropland (fig. 4). Fertilizer and animal livestock sources of nitrogen and phosphorus are relatively high in this basin and reflect the influence of the highly agricultural regions of the basin (figs. 5A–5B). Fertilizer inputs of N and P averaged 4.4 and 0.7 (g/m²)/yr during the period 1997–2001, respectively (fig. 5A). Animal livestock sources of N and P averaged 1.5 and 0.5 (g/m²)/yr, respectively (fig. 5B). Corn harvested in this drainage basin averaged over 10,000 bushels/km² during the period 2000–2010 and indicates fairly intense agricultural production throughout the entire basin (fig. 6).

There also has been extensive hydrologic modification along the main stem of the upper Mississippi River. The 1930 Flood Control Act mandated that the U.S. Army Corps of Engineers build and maintain a 9-ft navigational channel from St. Louis, Ill., to Minneapolis, Minn. This was accomplished primarily through the construction of a series of low-head navigational dams that alter the low-flow conditions, decrease contact between main channel and backwater habitats, and increase sedimentation rates behind the dams (Benke and Cushing, 2005).

Data Sources

Data sources for the main stem Upper Mississippi River are summarized in table 17. U.S. Geological Survey streamflow data are available from the reference stream gaging station at Grafton Ill., beginning in 1933. Water-quality data are available primarily from the Illinois State Water Survey and the U.S. Geological Survey. The Illinois State Water Survey began collecting data on rivers bordering or contained within Illinois beginning in 1895 and published these results regularly through 1908. Several sites included in this survey are on the upper Mississippi River including Alton, Ill., and Golden Eagle, Mo. Data from this survey are particularly notable because they include total nitrogen measurements, which are the most useful way to consider changes in nitrogen concentrations. The Illinois State Water Survey also collected water-quality data at several sites in the upper and middle Mississippi River from 1950 to 1975. U.S. Geological Survey data collection resumed in 1975 and between Grafton and Alton, Ill., water-quality data are available through 2009.

Middle Mississippi River

The middle Mississippi River is defined as the part of the river downstream of the confluence with the Missouri River and upstream of the confluence with the Ohio River (fig. 25). Although this stretch of river is only 300 km, it is important to consider separately because it integrates the influence of the upper Mississippi River with the Missouri River. Hydrologic differences between Alton and Thebes, Ill., reflect the influence of the Missouri River on the middle Mississippi River. For example, the total drainage area at the reference stream gaging station at Thebes, Ill. (station No. 07022000), is 1.85 million km² (714,000 mi²), mean annual discharge (1933–2008) is 5,900 m³/s (209,000 ft³/s), and runoff is 101 mm/yr. The Missouri River also is a major source of suspended sediments to the middle Mississippi River. More information about the Missouri River are provided in the Missouri River section.

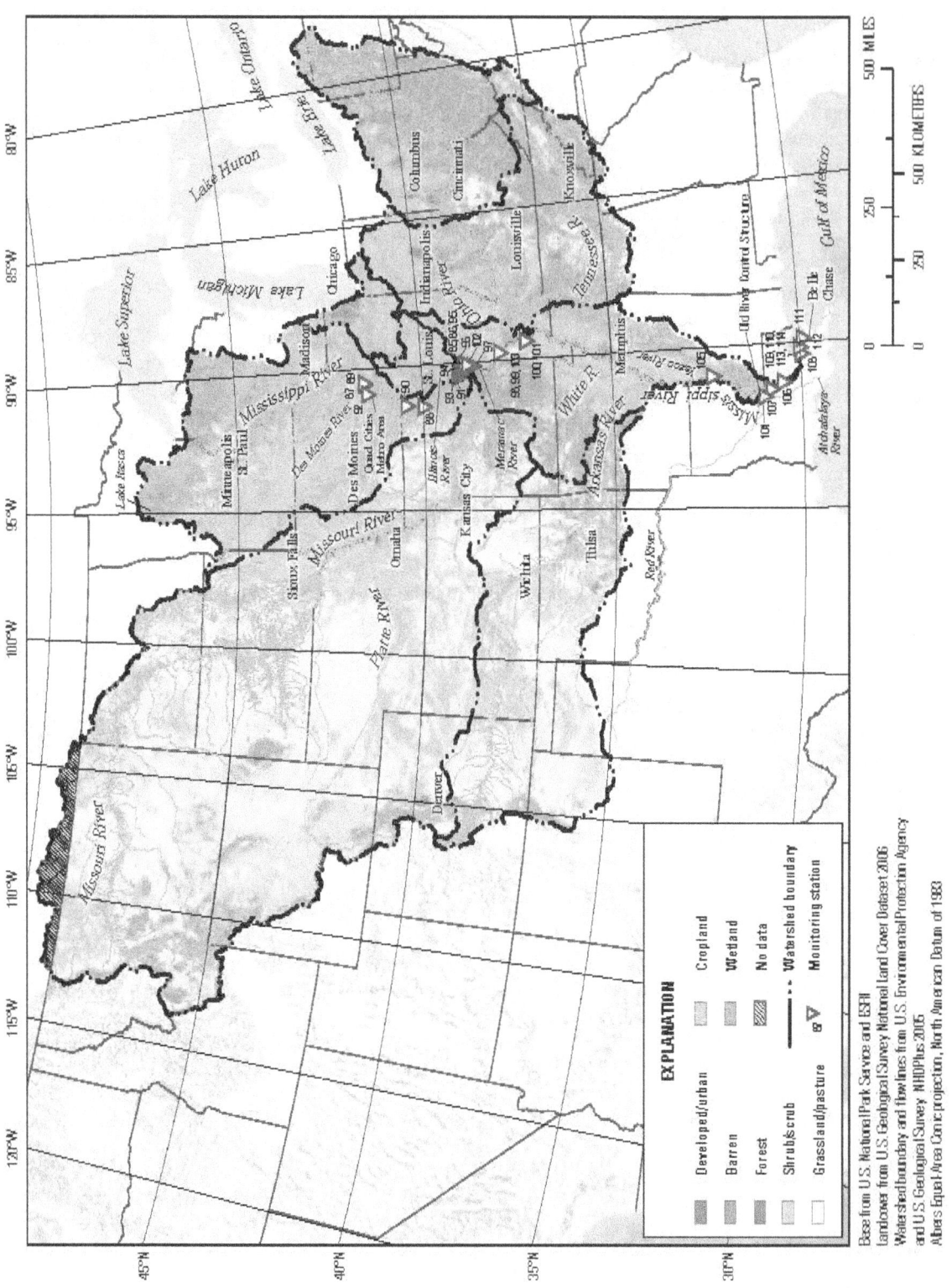

EXPLANATION

Developed/urban

Barren

Forest

Shrub/scrub

Grassland/pasture

Cropland

Wetland

No data

Watershed boundary

Monitoring station

Base from U.S. National Park Service and ESRI
landcover from U.S. Geological Survey National Land Cover Dataset 2006
Watershed boundary and flowlines from U.S. Environmental Protection Agency
and U.S. Geological Survey NHDPlus 2005
Albers Equal Area Conic projection, North American Datum of 1983

Figure 25. Mississippi River basin.

Human alteration of the middle Mississippi River channel dates back almost 200 years. After the passage of the Rivers and Harbors Act of 1824, the U.S. Army Corps of Engineers began to maintain a navigable 8-ft channel on the main stem Mississippi River to St. Louis. At first, this was achieved primarily through revetment and wing-dam construction but dredging became more prominent by the early 20th century. Levee construction to protect communities and farmland adjacent to the river also has been intense along this stretch of the river. Flood protection became a central focus after a large flood in 1903; most of the levee construction around St. Louis was completed by 1908 (Criss and Wilson, 2003). Since 1945, additional engineering has occurred in this region as flood-control levees were heightened and the entire main stem channel was restricted to 1,500 ft width (Criss and Wilson, 2003). Additionally, the Pick-Sloan Flood Control Act of 1944 authorized the construction of dozens of dams and reservoirs in the Missouri River basin, fundamentally altering streamflow in the Missouri River and dramatically reducing sediment loading to the middle Mississippi (Meade and Moody, 2010).

The middle Mississippi River also is distinguished from the upper Mississippi because of the proximity to St. Louis. To a large degree, the proximity to the Mississippi River has enabled St. Louis to remain a pivotal commercial center, so there has always been a close connection between the city and the river. St. Louis grew rapidly in the 19th century and by 1900 St. Louis was the fourth largest city in the country with a population approaching 600,000 (Criss and Wilson, 2003). Currently, almost 3 million people live in the 16 county metropolitan area making it the 18th largest in the United States (Criss and Wilson, 2003). St. Louis grew quickly before the advent of drinking water treatment or sewerage systems. The city has always drawn its drinking water from easily contaminated surface waters including the Missouri, Meramec, and Mississippi Rivers. Several early epidemics of waterborne illness resulted including a cholera outbreak in 1849, which claimed 5,000 lives (Criss and Wilson, 2003). Outbreaks of typhoid fever continued and in 1900 the city of St. Louis filed suit against the city of Chicago for discharging untreated sewage into the Illinois River, which resulted in degraded water quality in the Mississippi River downstream of the confluence with the Illinois River (Leighton, 1907). St. Louis slowly developed a sewer system, beginning in 1916; by 1956, only 5 percent of the population had wastewater treatment service (Criss and Wilson, 2003). In 1933, the city applied for a permit from the U.S. Army Corps of Engineers to dispose of solid waste garbage in the Mississippi River. Although the Bureau of Fisheries opposed the idea on the grounds that it would be detrimental to fish populations, the permit was issued in 1937 allowing the disposal of several hundred tons of solid waste garbage daily into the Mississippi River

(Ellis, 1943; Scarpino, 1985). Concerns about water quality grew throughout the middle 20th century and such practices became less common. By 1965, 95 percent of the St. Louis metropolitan area residents had wastewater treatment service and by 1972, the city became known as a leader in solutions to water-quality problems (Criss and Wilson, 2003).

Land use and population characteristics of the middle Mississippi River reflect the influence of its two primary tributaries – the upper Mississippi River and the Missouri River. Population density was 19 people/km^2 according to the 2000 U.S. Census (fig. 3). Almost 77 percent of the total basin area is in some agricultural use, with 40 percent in cropland (fig. 4). Other indicators of agricultural intensity were moderately high in this basin and reflect the joint influence of the intensively cultivated upper Mississippi River basin and Missouri River basin, which has less overall percentage of cropland area (fig. 4). Nitrogen from fertilizer and animal livestock sources averaged 3.5 g (N/m^2)/yr from 1997 to 2001 (fig. 5A). Phosphorus sources were approximately 0.7 g (P/m^2)/yr (fig. 5B). Corn harvest rates also were moderate at 4,100 bushels/km^2 of drainage area (fig. 6).

Data Sources

Data sources for the main stem Mississippi River are summarized in table 17. For the middle Mississippi River, intermittent streamflow data are available beginning in 1861 at St. Louis, Mo., with daily data available after 1928; daily streamflow data also are available beginning in 1933 at the reference stream gaging station at Thebes, Ill. (prior to 1941, published as "at Cape Girardeau, Mo."), and beginning in 1942 at Chester, Ill. (table 17). The Mississippi River at Chester, Ill., also was included in the Clarke (1924) compilation and sampled in 1906–07. Several sites on the middle Mississippi River were included as part of the Illinois State Water Survey water-quality sampling, including Thebes (1950–56), Chester, Ill. (1955–76), and East St. Louis, Ill. (1958–76). Regular water-quality sampling by the U.S. Geological Survey began in 1973 at Thebes, Ill.

Lower Mississippi River

The lower Mississippi River is defined as the approximately 1,500 km-long part of the river downstream of the confluence with the Ohio River at Cairo, Ill. (fig. 25). The Ohio River discharges 7,900 m^3/s (278,000 ft^3/s) annually (reference stream gaging station for lower Ohio, period 1928–2008) and approximately doubles the streamflow of the Mississippi River. Measured near Arkansas City, Ark., (USGS station 07265450; years 1928–1980) the mean annual discharge of the Mississippi River is 15,300 m^3/s

(540,000 ft³/s), the drainage area is 2.9 million km²
(1.1 million mi²), and mean annual runoff is 170 mm/yr. Other
significant tributaries to the lower Mississippi River include
the White, Arkansas, Red, and Yazoo Rivers, which deliver
another 5,700 m³/s (200,000 ft³/s) to the main stem river. The
Atchafalaya River is the main distributary of the Mississippi
River, which diverges from the main channel at river mile
315. Since 1963, the U.S. Army Corps of Engineers has
regulated the amount of water flowing into the Atchafalaya
River through the Old River Control Structure in Louisiana.
The purpose of this structure is to ensure that no more than
30 percent of the total flow of the Mississippi River enters the
Atchafalaya (Benke and Cushing, 2005).

The lower Mississippi River channel is located entirely
within the deep alluvial deposits of the Coastal Plain
physiographic province (fig. 2). The largest tributaries,
the Ohio, Arkansas, and Red Rivers, encompass many
physiographic provinces and land uses, and others originate
in the Ozark and Ouachita Plateaus (further details about the
Ohio and Arkansas Rivers are provided in respective sections
in this report).

Most of the entire length of the lower Mississippi River
channel is surrounded by levees. The floodplain inside the
levees is less than 10 percent of the total natural floodplain
area for this part of the river (Benke and Cushing, 2005). The
river channel itself has been extensively modified for flood
control including armoring shorelines, artificially installing
channel cutoffs, and building revetments to prevent channel
meandering. Therefore, the ecological integrity of this river-
floodplain ecosystem has been highly challenged.

Because the Mississippi River basin includes a large
portion of the continental United States, data to describe
population, land use, and agricultural activity encompass
many of the river basins included in this study. According to
the year 2000 census, population density was 25 people/km²
in the river basin considered as a whole, which was slightly
less than the national average of 31 people/km² (fig. 3). Just
more than 70 percent of the basin was agricultural land with

38 percent of the basin in cropland (fig. 4). Nitrogen from
fertilizer and animal livestock averaged 3.5 g (N/m²)/yr
from 1997 to 2001 with fertilizer sources being about twice
as large as animal livestock sources (fig. 5A). This amount
was virtually unchanged for several decades at the end of
the 20th century (fig. 18E). Phosphorus from these sources
totaled 0.6 g (P/m²)/yr and was split almost evenly between
fertilizer and animal livestock (fig. 5B). Corn harvest averaged
approximately 3,300 bushels/km² of drainage area (fig. 6).

Data Sources

Data sources for the main stem Mississippi River are
summarized in table 17. Daily streamflow data are available
for the reference stream gaging station at Tarbert's Landing,
Miss. beginning in 1930 (table 17). The earliest water-quality
data were collected by the Sewerage and Water Board of New
Orleans, La. in 1900–1901. The Mississippi River also was
sampled at New Orleans in 1906–07 and included as part of
the Clarke (1924) compilation. U.S. Geological Survey water-
quality sampling began in the 1950s at St. Francisville, Luling,
and New Orleans, La. The most current water-quality data,
beginning in the 1970s, are available at Belle Chasse, Baton
Rouge, and St. Francisville, La.

All water-quality and streamflow data for the lower
Mississippi River reach are located downstream of the
Atchafalaya River diversion and so only reflect water
quality and water discharge in that part of the stream, which
represents about 70 percent of the original water volume.
Therefore, analysis of these data should not be interpreted as
equivalent to estimates developed for the entire Mississippi-
Atchafalaya River basin. Furthermore, many different
water-quality and stream-gaging stations are included in our
summary of available data. The distance between some of
these stations is quite large. However, it is presumed that the
overall scale of the Mississippi River basin and the magnitude
of water carried by the main stream channel causes local
influences to be minimal and water-quality data between the
stations should be comparable.

Table 17.   Mississippi River basin data sources.

[**Monitoring station** is shown in figure 25. **Source:** NWIS, National Water Information System; STORET, Storage and Retrieval Data Warehouse; USACE, U.S. Army Corps of Engineers. **Constituent:** Alk, alkalinity. See table 2 for definitions of all other constituent abbreviations]

Monitoring station	Source	Station	Latitude	Longitude	Constituent	Start year	End year
			Upper Mississippi River				
[1]85	Long, 1889	Mississippi River at Alton, Ill.	38°53′06″	90°10′51″	NO$_3$	1888	1888
					NH$_3$	1888	1889
[1]86	Palmer, 1897, 1903; Bartow, 1907	Mississippi River at Alton, Ill.	38°52′55″	90°10′44″	TN	1896	1900
					NH$_3$	1896	1906
					TON + NH$_3$	1896	1900
					NO$_3$	1896	1906
					Alk	1904	1906
[1]87	Bartow, 1907, 1909	Mississippi River at Rock Island, Ill.	41°31′08″	90°34′01″	NH$_3$	1896	1908
					NO$_3$	1896	1908
					Alk	1896	1908
[1]88	Palmer, 1897, 1903; Bartow, 1907, 1909	Mississippi River at Qunicy, Ill.	39°56′30″	91°25′51″	TN	1895	1900
					NH$_3$	1895	1906
					TON + NH$_3$	1895	1900
					NO$_3$	1905	1908
					Alk	1905	1908
[1]89	Bartow, 1907, 1909	Mississippi River at Moline, Ill.	41°30′38″	90°31′06″	NH$_3$	1905	1907
					NO$_3$	1905	1907
					Alk	1905	1907
[1]90	Larson and Larson, 1957	Mississippi River at Keokuk, Iowa	40°23′27″	91°22′23″	NH$_3$	1950	1955
					NO$_3$	1950	1955
					Alk	1950	1955
[1]91	Palmer, 1897	Mississippi River at Golden Eagle, Mo.	38°52′18″	90°33′59″	TN	1896	1896
					NH$_3$	1896	1896
					TON + NH$_3$	1896	1896
[1]92	Wiebe, 1931	Mississippi River at Fairport, Iowa	41°26′07″	90°54′06′	NH$_3$	1929	1930
					NO$_3$	1929	1930
					TDP	1929	1930
[2]93	NWIS	05587450 Mississippi River at Grafton, Ill.	38°58′05″	90°25′04″	Daily streamflow	1933	2008
94	NWIS	05587455 Mississippi River below Grafton, Ill.	38°57′04″	90°22′16″	TN	1989	2009
					NH$_3$	1989	2009
					TON + NH$_3$	1989	2009
					NO$_3$	1989	2009
					Alk	1989	2009
					TP	1989	2009
					TDP	1989	2009
95	NWIS	05587500 Mississippi River at Alton, Ill.	38°53′06″	90°10′51″	Daily streamflow	1933	1987
96	NWIS	05587550 Mississippi River below Alton, Ill.	38°51′41″	90°08′15″	TN	1975	1989
					NH$_3$	1975	1989
					TON + NH$_3$	1975	1989
					NO$_3$	1975	1989
					Alk	1974	1989
					TP	1975	1989
					TDP	1975	1989

Table 17. Mississippi River basin data sources.—Continued

[**Monitoring station** is shown in figure 25. **Source:** NWIS, National Water Information System; STORET, Storage and Retrieval Data Warehouse; USACE, U.S. Army Corps of Engineers. **Constituent:** Alk, alkalinity. See table 2 for definitions of all other constituent abbreviations]

Monitoring station	Source	Station	Latitude	Longitude	Constituent	Start year	End year
			Middle Mississippi River				
97	NWIS	07010000 Mississippi River at St Louis, Mo.	38°37'44"	90°10'47"	Daily streamflow	1861	2008
[1]98	Long, 1889	Mississippi River at Chester, Ill.	37°54'14"	89°50'08"	NO$_3$	1888	1888
					NH$_3$	1888	1888
[1]99	NWIS; Clarke, 1924	07020500 Mississippi River near Chester, Ill.	37°54'14"	89°50'08"	Daily streamflow	1942	2008
					NO$_3$	1906	1907
					Alk	1906	1907
[2]100	NWIS	07022000 Mississippi River at Thebes, Ill.	37°12'59"	89°28'03"	Daily streamflow	1933	2008
					TN	1973	2008
					NH$_3$	1973	2008
					TON + NH$_3$	1973	2008
					NO$_3$	1973	2008
					Alk	1973	2008
					TP	1973	2008
					TDP	1973	2008
101	Larson and Larson, 1957	Mississippi River at Thebes, Ill.	37°12'59"	89°28'03"	NH$_3$	1950	1956
					NO$_3$	1950	1956
					Alk	1950	1956
102	Harmeson and Larson, 1969; Harmeson and others, 1973	Mississippi River at East St. Louis, Ill.	38°46'33"	90°09'35"	NH$_3$	1958	1976
					NO$_3$	1958	1976
					Alk	1958	1976
					TDP	1960	1976
103	Larson and Larson, 1957; Harmeson and Larson, 1969; Harmeson and others, 1973	Mississippi River at Chester, Ill.	37°54'14"	89°50'08"	NH$_3$	1955	1976
					NO$_3$	1955	1976
					Alk	1955	1976
					TDP	1960	1976
			Lower Mississippi River				
[2]104	USACE	01100 Mississippi River at Tarbert's Landing, Miss.	31°00'30"	91°37'25"	Daily streamflow	1930	2009
105	NWIS	07289000 Mississippi River at Vicksburg, Miss.	32°18'54"	90°54'21"	Daily streamflow	2008	2009
					TN	1973	1999
					NH$_3$	1975	1999
					TON + NH$_3$	1973	1999
					NO$_3$	1961	1999
					Alk	1961	1999
					TP	1973	1999
					TDP	1973	1999
106	NWIS	07374000 Mississippi River at Baton Rouge, La.	30°26'44"	91°11'30"	Daily streamflow	2004	2009
					TN	1975	2009
					NH$_3$	1975	2009
					TON + NH$_3$	1975	2009
					NO$_3$	1975	2009
					Alk	1975	2009
					TP	1975	2009
					TDP	1991	2009

Table 17. Mississippi River basin data sources.—Continued

[**Monitoring station** is shown in figure 25. **Source:** NWIS, National Water Information System; STORET, Storage and Retrieval Data Warehouse; USACE, U.S. Army Corps of Engineers. **Constituent:** Alk, alkalinity. See table 2 for definitions of all other constituent abbreviations]

Monitoring station	Source	Station	Latitude	Longitude	Constituent	Start year	End year
			Lower Mississippi River—Continued				
107	NWIS	07373420 Mississippi River near St. Francisville, La.	30°45′30″	91°23′45″	TN	1974	2009
					NH_3	1969	2009
					$TON + NH_3$	1974	2009
					NO_3	1954	2009
					Alk	1954	2009
					TP	1974	2009
					TDP	1973	2009
108	NWIS	07374400 Mississippi River at Luling, La.	29°56′19″	90°21′49″	TN	1974	1999
					NH_3	1991	1999
					$TON + NH_3$	1974	1999
					NO_3	1957	1999
					Alk	1957	1999
					TP	1973	1999
					TDP	1973	1999
109	NWIS	07374500 Mississippi River near New Orleans, La.	29°58′28″	90°10′14″	NO_3	1905	1955
					Alk	1905	1955
110	NWIS	07374508 Mississippi River at New Orleans, La.	29°57′03″	90°08′17″	TN	1973	1988
					NH_3	1974	1976
					$TON + NH_3$	1973	1988
					NO_3	1954	1988
					Alk	1954	1988
					TP	1973	1988
					TDP	1972	1984
111	NWIS	07374522 Mississippi River at Violet, La.	29°52′52″	89°54′02″	TN	1974	1976
					$TON + NH_3$	1974	1974
					NO_3	1973	1978
					Alk	1973	1973
					TP	1973	1978
					TDP	1973	1974
112	NWIS	07374525 Mississippi River at Belle Chasse, La.	29°51′25″	89°58′40″	Daily streamflow	2008	2009
					TN	1977	2009
					NH_3	1978	2009
					$TON + NH_3$	1977	2009
					NO_3	1977	2009
					Alk	1977	2009
					TP	1977	2009
					TDP	1978	2009
[1]113	STORET	210020 Mississippi River at New Orleans, La.	29°56′38″	90°10′07″	TN	1964	1971
					NH_3	1958	1969
					$TON + NH_3$	1964	1968
					NO_3	1964	1976
					Alk	1958	1976
					TP	1964	1976
					TDP	1964	1968
[1]114	Weston, 1903	Mississippi River at New Orleans, La.	29°56′40″	90°10′10″	NH_3	1900	1901
					NO_3	1900	1901
					Alk	1900	1901
					NH_3	1900	1901

[1]Historical station, location approximate.

[2]Reference stream gaging station.

Upper Colorado River Basins

The Colorado River is the largest river in the Southwestern United States, and drains a complex landscape ranging from high mountains to desert lowlands (fig. 26). The Colorado River supplies water for municipal and industrial use to about 27 million people, and water for irrigation of nearly 4 million acres of land (U.S. Department of the Interior, 2003). Water management in the Colorado River basin is intense and subject to the terms of the 1922 Colorado River Compact, which serves to divide the basin into upper and lower to facilitate the complex allocation of its water among seven Western States and Mexico. The division between the two basins is legally defined at the confluence of the Paria River, downstream of Glen Canyon Dam near Lees Ferry, Arizona. The total area for the basin upstream of Lees Ferry is about 290,000 km² (111,800 mi²). The upper Colorado region is the area of interest for this study, where the three major drainage basins include the main stem Colorado River, the Green River, and the San Juan River basins. Two sampling sites on the Colorado River were selected for this analysis: the Colorado River at Cisco, Utah, located just upstream of the confluence with the Green River; and the Colorado River at Lees Ferry, Ariz., representing the output from the upper basin. The Gunnison River also was included because it is the largest tributary to the upper Colorado River upstream of Cisco.

The upper Colorado River basin lies within the Colorado Plateau physiographic province (fig. 2) and is bounded by the Wyoming Basin to the north, the Middle and Southern Rocky Mountains to the east and south, and the Basin and Range to the west. As a consequence of the highly erodible character of much of the basin, the river has continually deepened its bed to create massive canyons in many areas through which the rivers flow (Bishop and Porcella, 1980). Climate patterns include short and warm summers and long, cold winters with deep snow in the mountains. Precipitation ranges widely, exceeding 125 cm (50 in.) at high elevations and reaching a minimum of about 15 cm (6 in.) in the high desert areas at lower elevation (Bishop and Porcella, 1980). Mean annual discharge (1921–2008) at the lowermost reference stream gaging station at Lees Ferry (station No. 09380000) is about 400 m³/s (14,000 ft³/s), indicating average runoff of about 43 mm/yr (among the lowest of streams in this study).

Because of the range of precipitation and topography, one of the key characteristics of the upper Colorado River basin is that the primary source of water is snowfall high in the mountains while the major course of the river itself and its tributaries runs through arid land. The area is subject to large-scale wet and dry climate cycles, which are associated with extreme variability in annual flow conditions that can range over an order of magnitude. Natural streamflow patterns were historically highly seasonal, with most of the runoff provided during the spring snowmelt period. To provide regulation of flow variability to meet downstream needs during summer low-flow periods, an extensive network of large dams was developed over the course of the 20th century. By 1980, more than 117 storage reservoirs had been constructed in the basin, primarily located in the mountainous areas of the upper reaches, supplying hydropower as well as supporting the export of pristine water out of the basin (Bishop and Porcella, 1980). Dozens of trans-mountain canals and tunnels transport water from these headwater regions to the South Platte, Arkansas, and Rio Grande basins (Bishop and Porcella, 1980).

The upper Colorado River basin is largely rural, having few population centers with more than 50,000 people. Population density for the entire upper basin is about 3 people/km², as determined by the 2000 census (fig. 3). Irrigated agriculture is the largest water use, although irrigated agricultural lands represent a small proportion of the total area (less than 5 percent in 1990) (U.S. Department of the Interior, 2003). Total agricultural area is low, just over 20 percent of the entire basin, and the primary agricultural land use is grazing. Irrigation water primarily is used to support livestock rather than crops (Benke and Cushing, 2005). Geology of the basin is dominated by calcium, sulfate, and bicarbonate dissolution ions (Stanford and Ward, 1991). As a result, although water-quality issues in the upper basin range from the impact of mining to reservoir effects, the most significant water-quality issue relates to the increasingly high levels of salinity that result from natural geologic sources concentrated by climatic factors and coincident evaporation from the many reservoirs, and anthropogenic water use.

The five basins considered in the Colorado River basin had the lowest population densities among all basins in this study, ranging from approximately 2 to 6 people/km² (fig. 3). These basins lack significant metropolitan areas so population growth was slow throughout the 20th century (fig. 27A). Agriculture, and especially cropland, is not a major land use in these basins, with less than 40 percent of their area in agricultural land and less than 7 percent in cropland in 2002 (fig. 4). In the mid-20th century, almost 60 percent of the San Juan River basin was reported as farmland in the Census of Agriculture (fig. 27C). However, among the other basins, agricultural land never surpassed 30 percent of basin area (fig. 27C). Nitrogen and phosphorus from fertilizer and animal livestock sources were among the lowest in this study averaging less than 0.6 g (N/m²)/yr and less than 0.15 g (P/m²)/yr, respectively, in these basins from 1997 to 2001 (figs. 5A–5B). Animal sources of nitrogen and phosphorus surpassed fertilizer sources in all basins over the same time period (figs. 5A–5B). Corn harvest averaged less than 100 bushels/km² in all basins of interest in the Colorado River with less than 1 bushel of corn/km² harvested on average between 2000 and 2010 in the San Juan basin (fig. 6). The relative magnitude of corn harvest intensity in these basins is emphasized by the fact that the lines are barely visible in figure 27B.

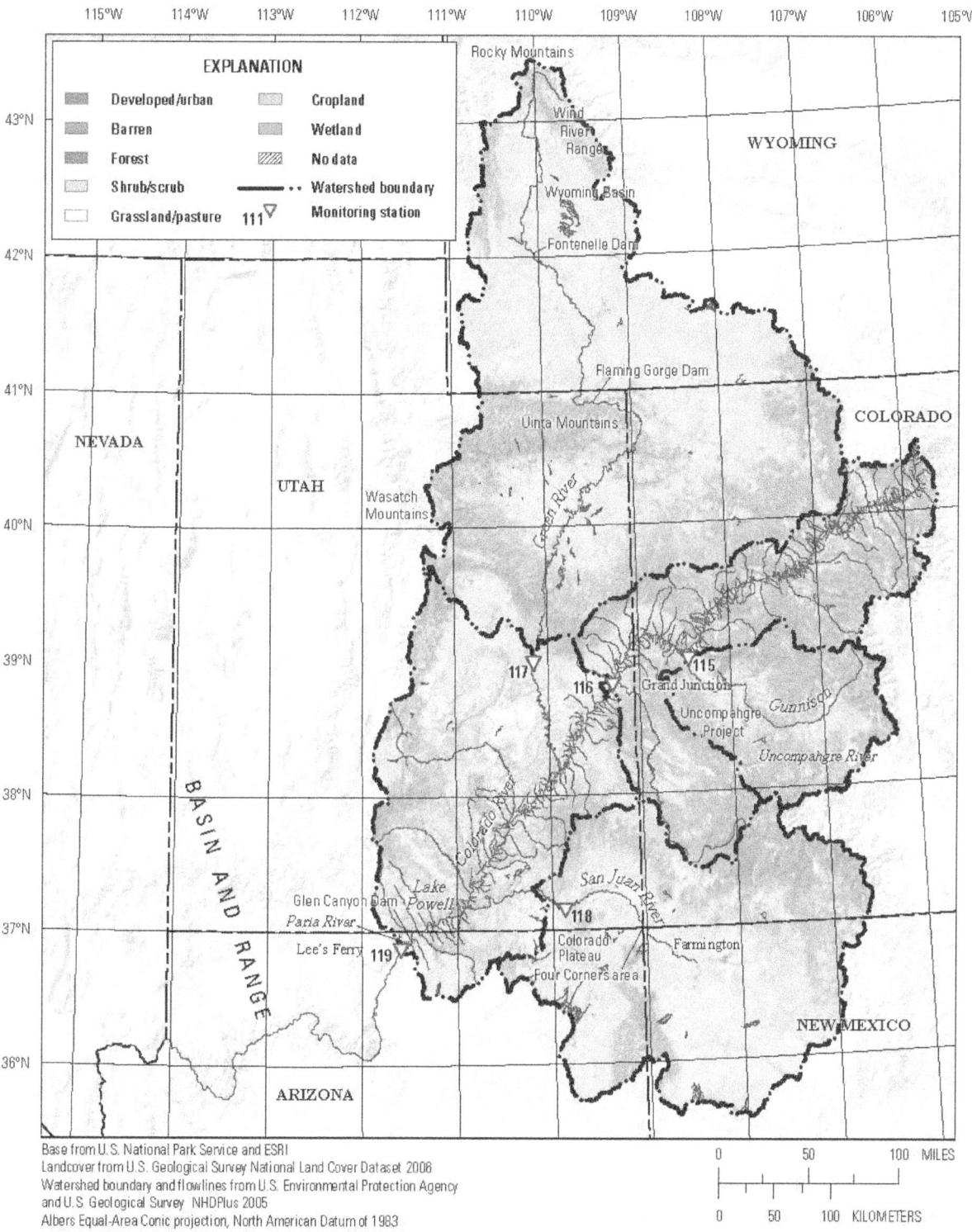

Figure 26. Upper Colorado River basins.

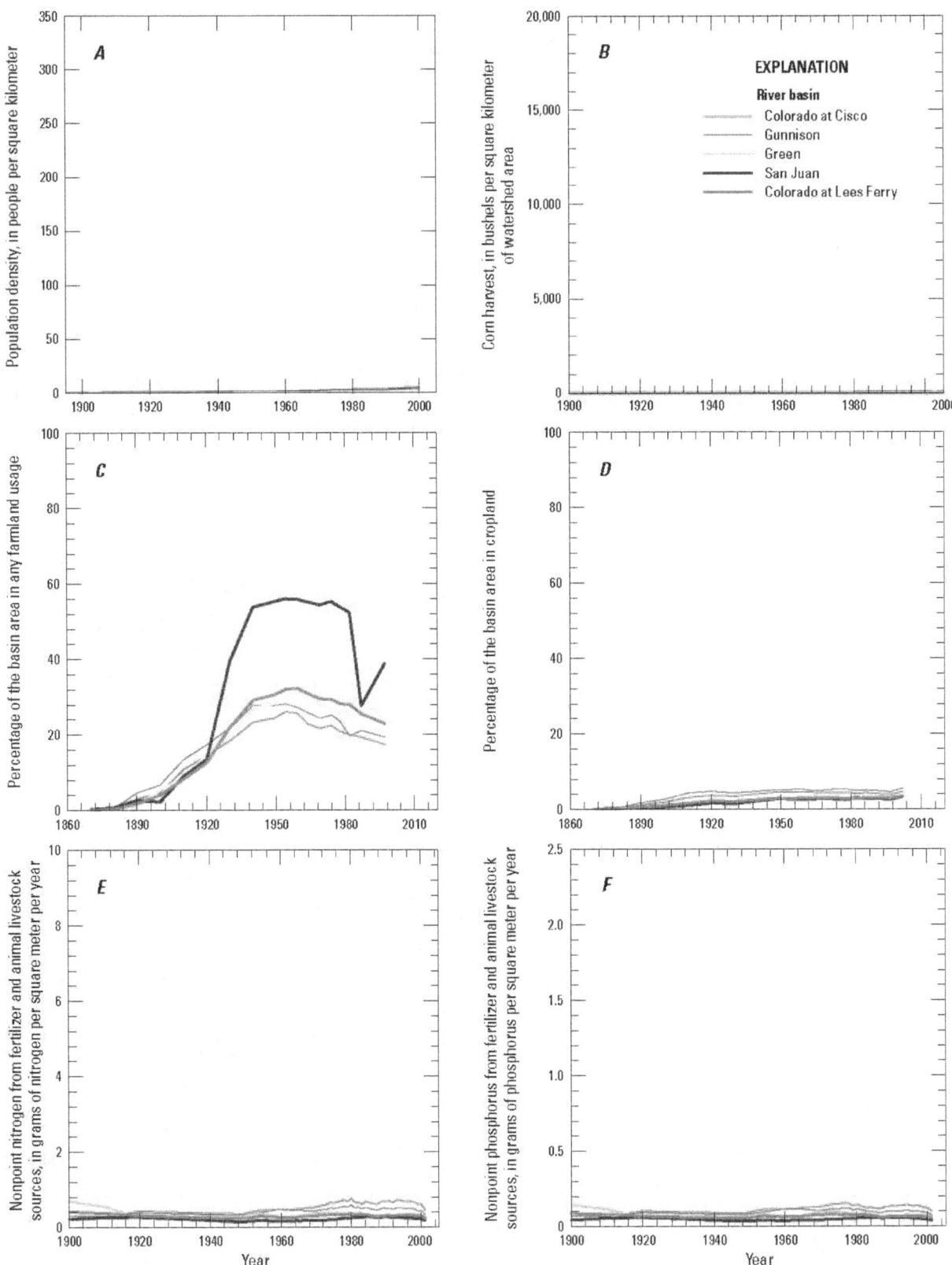

Figure 27.   Historical changes in (*A*) population, (*B*) corn harvest, (*C*) percentage of basin in any farmland usage, (*D*) percentage of basin in cropland, and fertilizer and animal livestock sources of (*E*) nitrogen and (*F*) phosphorus in the basins of interest in the Upper Colorado River.

Green River

The Green River drains the northernmost region of the upper Colorado basin, including areas in southwestern Wyoming, northeastern Utah, and northwestern Colorado (fig. 26). It is the largest tributary to the Colorado River, with a drainage area of approximately 116,000 km² (44,850 mi²) referenced to the USGS gage at Green River, Utah. Although the Green River originates in the Wind River Range in Wyoming, most of the river drains high desert dominated by dryland vegetation. Annual precipitation ranges from more than 100 cm (nearly 40 in.) in the mountains, falling mostly as snow, to less than 25 cm (about 10 in.) at low elevations (Benke and Cushing, 2005). Streamflow is regulated by a large number of dams, with the largest being the Fontanelle Dam in southwestern Wyoming and Flaming Gorge Dam in northeastern Utah. Mean annual discharge at the reference stream gaging station at Green River (station No. 09315000) (1894–2008) is about 175 m³/s (6,200 ft³/s), and runoff is 48 mm/yr.

The largest towns in the Green River basin are small (population less than 20,000) so urbanization is not a major factor for water quality. Population density was the lowest of all streams in this study, estimated as less than 2 people/km² in the 2000 U.S. Census (fig. 3). Although the major industry is mineral production focused on oil and natural gas (Bureau of Reclamation, 2010), agriculture is the most important land use in the basin (Benke and Cushing, 2005), with approximately 20 percent of the total basin area in some form of agricultural use in 2002 (fig. 4). Agriculture in the Green River basin is focused primarily on livestock production of beef, cattle, and sheep (Bureau of Reclamation, 2010), so that most water use in the basin is for irrigation of feed for livestock. Little fertilizer application is associated with these crops and so animal livestock was a much larger source of nitrogen and phosphorus in this basin (figs. 5A–5B), although these sources are among the lowest in this study.

Colorado River at Cisco

The part of the Colorado River basin upstream of the confluence with the Green River covers about 62,400 km² (24,000 mi²), as referenced by the USGS gage at Cisco, Utah (fig. 26). The gradient in topography and precipitation is similar to the Green River basin, with most precipitation falling in the eastern and southern mountains as snow and very little precipitation associated with the lower elevations farther west (Apodaca and others, 1996). Despite the smaller area of this subbasin compared to the Green River basin, mean

annual discharge for the reference stream gaging station at Cisco (station No. 09180500) (1913–2008) is higher at about 210 m³/s (7,400 ft³/s), and runoff is correspondingly higher as well (106 mm/yr).

The largest town in the sub-basin is Grand Junction, Colo., which has a population of about 60,000 (U.S. Census Bureau, 2000), although most towns have populations less than 10,000 (Apodaca and others, 1996). Population density was correspondingly low at the end of the 20th century, approximately 5.5 people/km² according to the 2000 census (fig. 3). Mineral production is the predominant industry, including gold in the past and currently focused on molybdenum (Bureau of Reclamation, 2010). Most land use is designated as rangeland or forest, with large areas set aside as National and State parks, wilderness areas, and ski areas; irrigated agricultural lands are limited to river valleys and low altitude areas. Irrigation accounts for about one-half of water use, generally devoted to livestock feed but also including crops, such as corn, beans, vegetables, and fruit (Bureau of Reclamation, 2010). As with the other basins of interest in this region, nitrogen and phosphorus sources from fertilizer and animal livestock were extremely low (figs. 5A–5B).

Gunnison River

The Gunnison River is the largest tributary to the upper Colorado River, draining an area of about 20,000 km² (nearly 8,000 mi²) in western Colorado (fig. 26) with a mean annual discharge at the reference stream gaging station near Grand Junction (station No. 09152500) (1896–2008) of about 72 m³/s (2,500 ft³/s). Elevation ranges from more than 4,000 m (about 13,800 ft) to approximately 1,400 m (4,500 ft) (Williams and others, 2009). Population density was very low, about 4 people/km² in the 2000 census (fig. 3). The basin encompasses two physiographic provinces: the Southern Rocky Mountains in the headwaters and Canyonlands subprovince of the Colorado Plateau in the lower reaches (fig. 2; Apodaca and others, 1996). Mean annual runoff (1896–2008) is relatively high at 110 mm, reflecting the largely mountainous character of the drainage basin. One of the first Federal reclamation projects in the Western United States was in the Gunnison basin: the Uncompahgre Project, which has provided irrigation water to lands in the Uncompahgre River valley since 1909 (Butler and others, 1991). Like other basins in the upper Colorado River, agricultural land primarily was non-cropland and comprised around 20 percent of total basin area in 2002 (fig. 4). Nonetheless, irrigation has been implicated in increased levels of dissolved solids and nitrate in the Gunnison River (Butler and others, 1991).

San Juan River

The San Juan River at Bluff, Utah, drains an area of nearly 60,000 km² (23,000 mi²) in the Four Corners area of Arizona, Colorado, Utah, and New Mexico (fig. 26). It is the second largest tributary to the Colorado River and empties into the middle of Lake Powell, which is the reservoir behind Glen Canyon Dam. Climate and topography are similar to that of the upper basin as a whole, ranging from alpine mountains dominated by snowmelt to desert lowlands that contribute little runoff to the river. Mean annual discharge for the reference stream gaging station at Bluff, Ut. (station No. 09379500) for the period 1914–2008 is 71 m³/s (2,500 ft³/s); annual runoff is estimated as 36 mm/yr, the lowest of all streams considered in this study.

Land ownership is dominated by Native American reservations, which cover approximately 60 percent of the San Juan River basin (Kirkpatrick, 2000). National forests and parks also represent a significant component, with only about 13 percent of the basin held in private ownership (Kirkpatrick, 2000). Energy and mineral extraction are important industries in the San Juan River basin, especially natural gas, crude oil, and coal (Kirkpatrick, 2000). Agriculture is equally important as a component of the local economy, with most cropland focused on feed for livestock although corn, dry beans, truck gardens, and orchards are cultivated in low-elevation areas close to the river (Bureau of Reclamation, 2010). Irrigation accounts for more than 90 percent of water usage (Bureau of Reclamation, 2010).

Colorado River at Lees Ferry

The Colorado River at Lees Ferry represents outflow from Lake Powell, a large hydroelectric and storage reservoir that fills the canyon behind Glen Canyon Dam. Glen Canyon Dam was closed in 1963 and the reservoir reached full pool 20 years later in 1983 (Stanford and Ward, 1991). As a result, water quality in the Colorado River at this site since that time primarily reflects the limnological processes occurring in the reservoir, interacting with the operation of the dam. Lake Powell is the second largest reservoir in the United States, with a maximum depth of 171 m and volume exceeding 33 km³ (Stanford and Ward, 1991). The water budget of the reservoir is largely determined by the legal allocation of water between the upper and lower basins, based on the Colorado River Compact of 1922. Inflow primarily is from the Colorado and San Juan Rivers and occurs during snowmelt in late spring and early summer. Because the average residence time for water is about 1.2 years, the reservoir acts as a critical buffer between the upper and lower basins when inflow is reduced (Gloss and others, 1981). The reservoir has a significant effect on downstream water quality.

Streamflow patterns have significantly changed in the Colorado River downstream of Lake Powell, and are now essentially determined by the needs of hydropower production (Stanford and Ward, 1991). Water chemistry in the river at Lees Ferry reflects the intensity of seasonal stratification occurring in the reservoir, which is driven largely by interaction of inflowing water with cold and saline water on the bottom of the reservoir, resulting in overflowing or interflowing density currents (Stanford and Ward, 1991). Stratification is further enhanced by the reduction of wind-driven circulation due to the preponderance of vertical cliffs along the shoreline, so that the reservoir never fully mixes to the bottom (Johnson and Merritt, 1979). Nutrient concentrations in the mixed water near the surface are strongly determined by phytoplankton production, being more uniform during the winter and early spring and declining during the summer growing season (Stanford and Ward, 1991). Water delivered during the spring flood tends to remain near the surface of the reservoir, so that inflowing nutrients are readily taken up by phytoplankton and removed from the system (Gloss and others, 1981). Water withdrawal is focused in the metalimnion. The reservoir essentially functions as a nutrient sink and retains more than 80 percent of the incoming nutrient load (Stanford and Ward, 1991). Phosphorus removal is especially significant, mediated by calcite precipitation as primary production increases over the summer, and reflected in near total retention of phosphorus in the reservoir (Stanford and Ward, 1991).

Data Sources

Data sources for the upper Colorado River basins are summarized in table 18. Daily streamflow data have been collected by USGS at all sites for all or most of the 20th century (Green River at Green River, Utah, 09315000, since 1894; Colorado River at Cisco, 09180500, since 1913; Gunnison River near Grand Junction, Colo., 09152500, since 1896; San Juan River near Bluff, Utah, 09379500, since 1914; Colorado River at Lees Ferry, Ariz., 09380000, since 1921. Water-quality data, including major ions and nitrate, from USGS are available beginning in the 1920s (table 18).

Table 18.   Upper Colorado River basin data sources.

[**Monitoring station** is shown in figure 26. **Source:** NWIS, National Water Information System. **Constituent:** Alk, alkalinity. See table 2 for definitions of all other constituent abbreviations]

Monitoring station	Source	Station	Latitude	Longitude	Constituent	Start year	End year
[1]115	NWIS	09152500 Gunnison River near Grand Junction, Colo.	38°59′00″	108°27′00″	Daily streamflow	1896	2008
					TN	1975	1998
					NH_3	1977	2002
					$TON + NH_3$	1975	1998
					NO_3	1931	2002
					Alk	1931	2009
					TP	1975	2002
					TDP	1967	2002
[1]116	NWIS	09180500 Colorado River at Cisco, Utah	38°48′38″	109°17′34″	Daily streamflow	1913	2008
					TN	1974	2000
					NH_3	1977	2000
					$TON + NH_3$	1974	2000
					NO_3	1928	2000
					Alk	1928	2008
					TP	1974	2000
					TDP	1969	2000
[1]117	NWIS	09315000 Green River at Green River, Utah	38°59′10″	110°09′04″	Daily streamflow	1894	2008
					TN	1974	2000
					NH_3	1977	2000
					$TON + NH_3$	1974	2000
					NO_3	1928	2000
					Alk	1928	2000
					TP	1974	2000
					TDP	1968	2000
[1]118	NWIS	09379500 San Juan River at Bluff, Utah	37°08′49″	109°51′53″	Daily streamflow	1914	2008
					TN	1974	2000
					NH_3	1977	2000
					$TON + NH_3$	1974	2000
					NO_3	1928	2000
					Alk	1928	2009
					TP	1974	2000
					TDP	1968	2000
[1]119	NWIS	09380000 Colorado River at Lee's Ferry, Ariz.	36°51′53″	111°35′15″	Daily streamflow	1921	2008
					TN	1974	2008
					NH_3	1977	2008
					$TON + NH_3$	1974	2008
					NO_3	1926	2008
					Alk	1926	2008
					TP	1974	2008
					TDP	1971	2000

[1]Reference stream gaging station.

Western Basins

Western basins include the Willamette River in the Pacific Northwest region, and the San Joaquin River and the Santa Ana River in the California region (fig. 1). The Santa Ana River basin was the most densely populated basin in this study, with almost 330 people/km² (fig. 3). In contrast to other densely populated basins in the east, population densities were low or moderate in the Santa Ana River basin until the 1940s (fig. 28A). After then, population has grown tremendously, surpassing the Schuylkill River basin by the year 2000 (fig. 3). Population growth in the Santa Ana River basin was greatly augmented by importation of water from the Colorado River beginning in 1928 and by the presence of a large military training facility used during World War II, which catalyzed population growth after cessation of hostilities (further details provided in the Santa Ana River section).

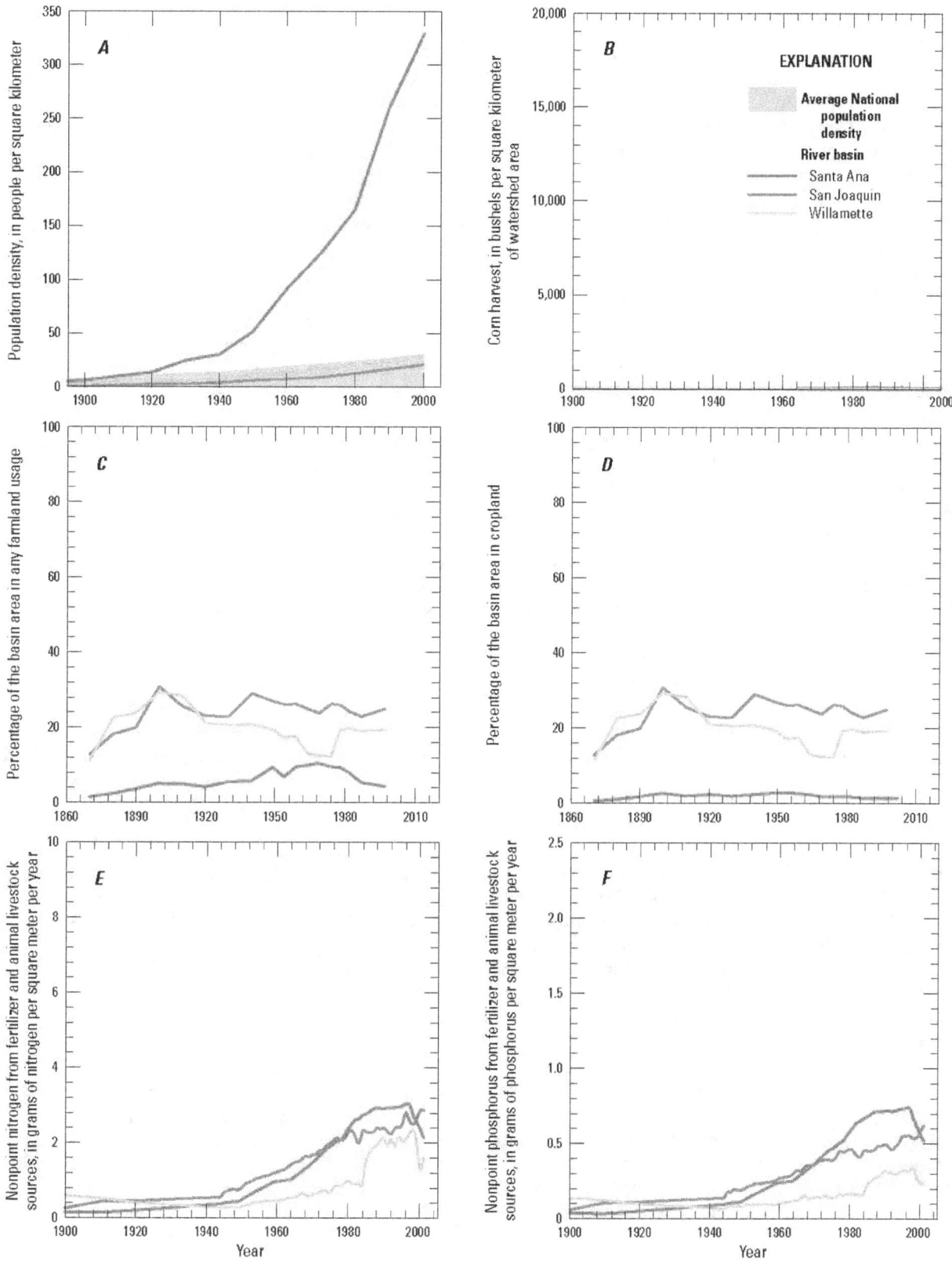

Figure 28. Historical changes in (A) population, (B) corn harvest, (C) percentage of basin in any farmland usage, (D) percentage of basin in cropland, and fertilizer and animal livestock sources of (E) nitrogen and (F) phosphorus in the basins of interest in the Western United States.

Santa Ana River

The Santa Ana River is one of the largest rivers in the Los Angeles region of southern California, although it drains a relatively small basin (fig. 29). The river's headwaters arise high in the San Bernardino Mountains, from which it flows roughly west through the Santa Ana Mountains down onto the Orange County coastal plain before discharging into the Pacific Ocean (California Regional Water Quality Control Board, 1995). The topography of the basin is diverse despite the small size, including mountain peaks up to 3,500 m (11,500 ft) above sea level, broad inland alluvial valleys, and the coastal plain in the west. The basin includes two ecoregions: the California Montane Chaparral and Woodlands and the California Coastal Chaparral and Woodlands (Benke and Cushing, 2005) of the Lower Californian physiographic province (fig. 2). The area of interest for this study is the upper part of the drainage area, located upstream of the Chino Hills and the Santa Ana Mountains, with a drainage area of about 5,800 km² (2,260 mi²) as referenced by the USGS gage below Prado Dam. Located within the basin is the closed San Jacinto River/Lake Elsinore drainage, which covers 3,860 km² (1,500 mi²) and very rarely provides discharge to the Santa Ana River—only several times during the 20th century. This area is thus considered non-contributing to downstream flow (Burton and others, 1998). This area of southern California is currently highly urbanized and densely populated. The Santa Ana River basin is the most densely populated basin included in this study, 330 people/km², according to the 2000 census (fig. 3).

Climate in the Santa Ana River basin is arid to semi-arid; summer is warm and dry while most precipitation falls during the relatively mild winter (November–March). Average annual precipitation is about 38 cm (15 in.) (California Regional Water Quality Control Board, 1995), with a slight gradient that ranges from lower near the coast to higher in the inland valleys and in the mountains (Izbicki and others, 2000). Mean annual discharge (USGS station 11074000; 1940–2008) is about 5 m³/s (187 ft³/s), indicating runoff of about 26 mm/yr if the entire basin, including the non-contributing area, is considered in the calculation. Active exchange between surface water and large reservoirs of groundwater were characteristic of the river prior to significant development and use of water resources in the basin. Streamflow in the river has ebbed and flowed with a complementary pattern of groundwater recharge and discharge, responding to the presence of large geologic faults that forced water to the surface (California Regional Water Quality Control Board, 1995). Downstream of Prado Dam, the Santa Ana River is now routed to infiltration ponds in order to recharge the underlying aquifers of the coastal plain. Water quality in the river is of great interest because these aquifers provide the primary source of supply for about 2 million people in Orange County, California (Izbicki and others, 2000).

Although the Santa Ana River basin has been home to people for many centuries, settlement by Europeans began during the 18th century when the Spanish arrived. Development of the floodplain began during the 19th century when settlers started diverting water from the river for irrigation of their gardens. Large-scale agriculture became more prominent as the population grew, and irrigation was a key factor supporting the growth of citrus orchards and vineyards. By 1928, surface waters were at risk throughout the region and the Metropolitan Water District of Southern California was formed to import water from the Colorado River to southern California, including the Santa Ana River. Initially, little water was imported because most demand was being met by widespread pumping of groundwater. Nonetheless, increased demand for water in the Los Angeles area in the middle of the century was an important impetus for the development of the California State Water Project, which was approved in 1960 to transport water from northern to southern California, further augmenting streamflow in the Santa Ana River. Groundwater pumping has remained the largest source of water to the basin, providing about two-thirds of the total water demand by the end of the 20th century while imported water provided about one-quarter (Kratzer and others, 2010).

During World War II, the U.S. Army Air Corps established a training center in Santa Ana, which contributed to the large population growth that occurred in the basin after the war ended. This population growth was further stimulated by the proliferation of the entertainment, tourism, and aircraft industries in the Los Angeles area through the second half of the 20th century. During this time, population in the Santa Ana River basin increased significantly (California Regional Water Quality Control Board, 1990). Accordingly, the balance of land use has shifted over the course of the 20th century from primarily agricultural to urban. About 75 percent of water use was estimated as urban in the basin by the end of the century, with the remaining 25 percent estimated for agricultural use (Kratzer and others, 2010). As a consequence of this intense urban water use, by late in the 20th century, base flow in the river had become dominated by treated effluent from wastewater-treatment plants (Izbicki and others, 2000).

Among river basins considered in this study, the Santa Ana River basin had the lowest proportion used for farmland at 4 percent, with only 1 percent in cropland, in 2002 (fig. 4). Nonetheless, extensive dairy operations are concentrated in the Chino Valley and may affect nutrient inputs to the Santa Ana River. The Chino basin provides rising groundwater discharge to the Santa Ana River near Prado Dam, estimated as accounting for 5 to 10 percent of base flow (California

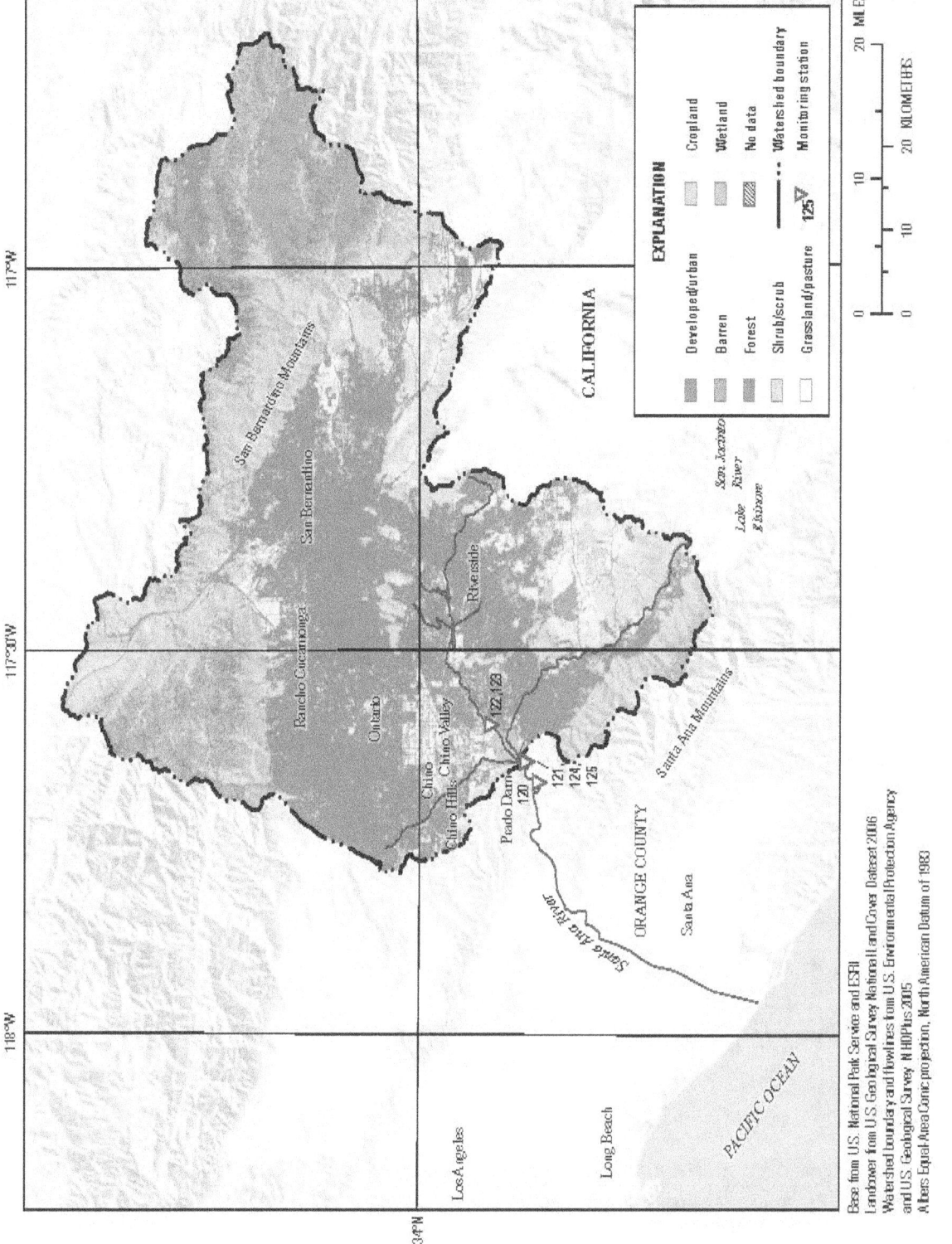

EXPLANATION

Developed/urban	Cropland
Barren	Wetland
Forest	No data
Shrub/scrub	— · — Watershed boundary
Grassland/pasture	125 ▽ Monitoring station

CALIFORNIA

San Bernardino Mountains

San Bernardino

Rancho Cucamonga

Riverside

Ontario

Chino   Chino Valley

Chino Hills

Prado Dam

120

121
124
125

122,123

Santa Ana Mountains

San Jacinto River

Lake Elsinore

ORANGE COUNTY

Santa Ana

Santa Ana River

Los Angeles

Long Beach

PACIFIC OCEAN

0   10   20 MILES

0   10   20 KILOMETERS

Base from U.S. National Park Service and ESRI
Landcover from U.S. Geological Survey National Land Cover Dataset 2006
Watershed boundary and flowlines from U.S. Environmental Protection Agency
and U.S. Geological Survey NHDPlus 2005
Albers Equal-Area Conic projection, North American Datum of 1983

Figure 29.   Santa Ana River basin.

Regional Water Quality Control Board, 1990). This groundwater contains significant loads of nitrogen, calculated to be about 30–40 percent of the nitrate measured in the river at the Prado Dam. These nitrogen loads originate largely from the discharge of wastewater from the high density of dairies located in the Chino Valley (California Regional Water Quality Control Board, 1990). Dairies began to be concentrated in this area during the 1970s; approximately 300 dairies supporting nearly 300,000 dairy cows were reported in 1995 (Burton and others, 1998). The Chino basin represents an area of about 650 km^2 (250 mi^2), and this density is considered among the highest concentration of dairy cows in the world (California Regional Water Quality Control Board, 1990).

Data Sources

Data sources for the Santa Ana River basin are summarized in table 19. USGS streamflow data are available for the reference stream gaging station below Prado Dam (station No. 11074500) beginning in 1919, and through the 20th century including data from a second station below Prado Dam (station No. 11074000). Water-quality data for the Santa Ana River near Corona, California, from 1906 to 1907 are available from the Clarke (1924) report, including major ions and nitrate. Inorganic carbon and nitrate data were collected by the California Department of Water Resources below Prado Dam beginning in 1950. The USGS began sampling for nitrate and alkalinity below Prado Dam in 1966 and continued through 2008.

Table 19. Santa Ana River basin data sources.

[**Monitoring station** is shown in figure 29. **Source:** NWIS, National Water Information System; STORET, Storage and Retrieval Data Warehouse. **Constituent:** Alk, alkalinity. See table 2 for definitions of all other constituent abbreviations]

Monitoring station	Source	Station	Latitude	Longitude	Constituent	Start year	End year
120	NWIS	11074500 Santa Ana River at County Line below Prado Dam, Calif.	33°52′24″	117°40′17″	Daily streamflow	1919	1960
[1]121	NWIS	11074000 Santa Ana River below Prado Dam, Calif.	33°53′00″	117°38′40″	Daily streamflow	1940	2008
					TN	1970	2008
					NH$_3$	1971	2008
					TON+NH$_3$	1970	2008
					NO$_3$	1966	2008
					Alk	1966	2008
					TP	1971	2008
					TDP	1970	2008
122	NWIS	11068000 Santa Ana River at River Road near Corona, Calif.	33°55′21″	117°35′46″	NH$_3$	1997	1997
					NO$_3$	1966	1997
					Alk	1968	1997
					TDP	1997	1997
[2]123	Clarke, 1924	Santa Ana River near Corona, Calif.	33°55′21″	117°35′46″	NO$_3$	1906	1907
					Alk	1906	1907
124	STORET	Y1155000 Santa Ana River below Prado Dam, Calif.	33°53′03″	117°35′42″	TN	1975	1978
					NH$_3$	1973	1979
					TON+NH$_3$	1975	1978
					NO$_3$	1950	1988
					Alk	1950	1988
					TP	1973	1979
					TDP	1974	1974
125	STORET	WB08Y1155000 Santa Ana River below Prado Dam, Calif.	33°53′03″	117°38′42″	TN	1975	1978
					NH$_3$	1973	1978
					TON+NH$_3$	1975	1978
					NO$_3$	1950	1978
					Alk	1950	1978
					TP	1973	1978
					TDP	1974	1974

[1]Reference stream gaging station.

[2]Historical station, location approximate.

San Joaquin River

The San Joaquin River is the second longest river in California, extending for a total length of 560 km (about 350 mi) to drain the southern two-thirds of the Central Valley (fig. 30). The major area of the basin includes parts of three physiographic regions: the Coast Ranges, San Joaquin Valley, and the Sierra Nevada (Kratzer and others, 2011) of the Cascade-Sierra Mountains and Pacific Border physiographic provinces (fig. 2). The broad and flat expanse of the San Joaquin Valley is bounded on the east by the Sierra Nevada Mountains, the south by the Tehachapi Mountains, which are located approximately 190 miles south of Merced, CA, and on the west by the Coast Ranges, while the Sacramento–San Joaquin Delta defines the northern boundary. The southern part of the basin is essentially hydrologically separate from the northern part and characterized by the closed or endorheic Tulare basin, which is located approximately 170 miles southeast of the Sacramento-San Joaquin Delta, that flows into the San Joaquin River only under exceptionally wet conditions. For this study, the basin area is defined as the northern perennial region with the river mouth at Vernalis, and equals 19,150 km² (about 7,400 mi²) (Kratzer and others, 2011). Major tributaries include the Stanislaus, Tuolumne, and Merced Rivers, which originate in the Sierra Nevada. Streams draining the Coast Ranges in the west are smaller and generally intermittent.

Climate in the San Joaquin basin is considered arid to semi-arid, with hot summers and relatively mild winters. A strong gradient of precipitation occurs across the basin from east to west, with little precipitation falling in the western region because of the rain shadow of the Coast Ranges. In contrast, the eastern part of the basin receives heavy precipitation on the western slopes of the Sierra Nevada that occurs as both rainfall and snow, and provides essentially all runoff to the river. Mean annual discharge (1923–2008) leaving the basin at the reference stream gaging station at Vernalis (station No. 11303500) is about 127 m³/s (4,500 ft³/s) (Kratzer and others, 2011), which indicates annual water yield of 210 mm. Seasonal patterns of streamflow were originally heavily influenced by snowmelt peaks in the spring and early summer, with lowest flow magnitudes occurring late in the summer and early autumn. These patterns have been significantly altered by reservoir storage in the second half of the 20th century, which generally acts to equalize flow volume throughout the year.

Over the course of the 19th and 20th centuries, an extensive network of channels and diversions has been built throughout the basin to divert surface water for irrigation. Originally, these channels were hand-dug and therefore agriculture was restricted to land close to the surface water supply. After the discovery of gold in the Serra Nevada in the mid-19th century, the development of water resources greatly expanded to meet the needs of the miners and the growing population. This development included construction of thousands of miles of diversion ditches and flumes for sluicing gold, which were later converted to irrigation use after mining ceased. By early in the 20th century, nearly all surface-water supply in the San Joaquin Valley had been diverted to support irrigation of agriculture (Gronberg and others, 1998). Subsequent development in the mid-20th century included large-scale Federal and State projects that import water to the basin from the Sacramento River and Trinity River, which is located approximately 200 miles north of the Sacramento-San Joaquin Delta, in the north via the Delta–Mendota Canal and the California Aqueduct. Additionally, streamflow in most of the major streams in the Sierra Nevada is currently controlled by reservoirs, which are largely managed for irrigation and hydropower production. Most water use in the basin is focused on agriculture, although surface water also is supplied to municipal users in southern California and the San Francisco area (Gronberg and other, 1998).

Because most of the water in the San Joaquin River originates in the Sierra Nevada, logging and mining land use there has important influences on water quality. Additionally, several cities, and other wastewater sources in the valley, discharge directly to the river (Kratzer and Shelton, 1998). Nonetheless, the greatest source of water-quality degradation in the river is agricultural water use, either due to direct return flow of runoff from irrigated fields or inflow from subsurface tile drain systems. Because irrigation water serves as a primary source of recharge to the groundwater system, irrigation return flow also can be routed to the river as groundwater discharge. Major crops grown include fruit and nuts, cotton, vegetables, and some grains, and livestock also is an important component of agricultural land use in the basin (Gronberg and others, 1998).

In 2002, although only 25 percent of the San Joaquin basin was agricultural (fig. 4), virtually the entire valley floor was in cropland or rangeland (fig. 30). Therefore, agricultural influence on water quality in this basin may not be most accurately described using basin-wide estimates of ancillary data. Also, in contrast to the intensively cultivated corn belt river basins such as the Maumee, Illinois, and Des Moines Rivers (figs. 4, 5A–5B, and 6), the focus of agricultural production in the San Joaquin Valley is not on corn agriculture. Thus, the corn harvest rate in this basin was very low, averaging 52 bushels/km² from 2000 to 2010, which was less than 1 percent as large as in the corn belt river basins (fig. 6). Fertilizer and animal livestock sources of nitrogen and phosphorus were the highest of the Western river basins (figs. 5A–5B). Population density was low, 21 people/km² (fig. 3).

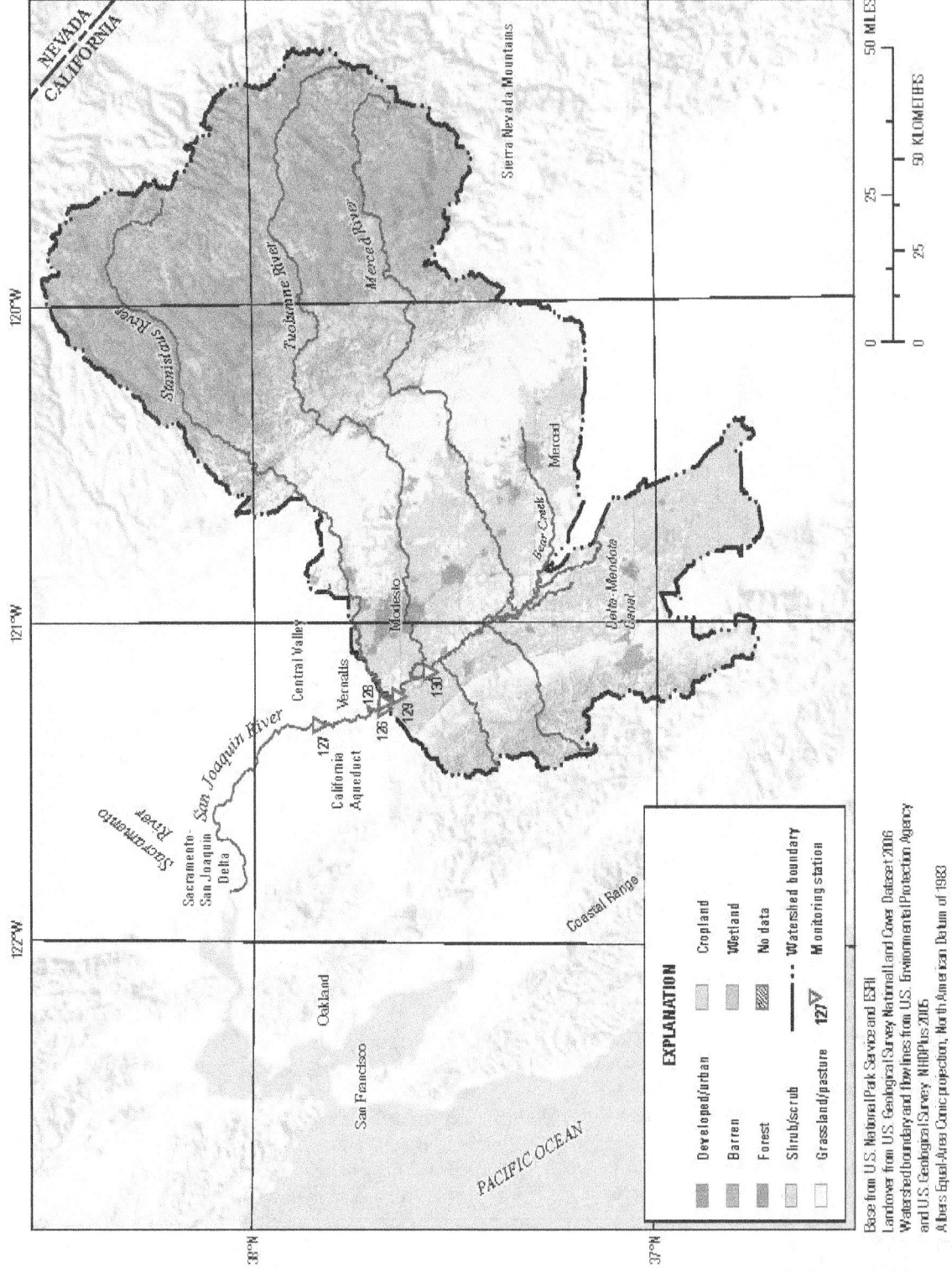

Figure 30.   San Joaquin River basin.

Data Sources

Data sources for the San Joaquin River basin are summarized in table 20. Daily streamflow data have been collected by USGS at the reference stream gaging station at Vernalis since 1923, except for 1925–28. Water–quality data for the San Joaquin River at Lathrop, California, from 1907 to 1908, are available from Clarke (1924), including major ions and nitrate. Inorganic carbon data were collected at Vernalis by the Bureau of Reclamation beginning in the 1930s through the 1980s. Additional data were collected by USGS and the California Department of Water Resources beginning briefly in the 1930s, and continuing in the 1950s through the second half of the 20th century.

Table 20.   San Joaquin River basin data sources.

[**Monitoring station** is shown in figure 30. **Source:** NWIS, National Water Information System; STORET, Storage and Retrieval Data Warehouse. **Constituent:** Alk, alkalinity. See table 2 for definitions of all other constituent abbreviations]

Monitoring station	Source	Station	Latitude	Longitude	Constituent	Start year	End year
[1]126	NWIS	11303500 San Joaquin River near Vernalis, Calif.	37°40'34"	121°15'55"	Daily streamflow	1923	2008
					TN	1973	2008
					NH_3	1977	2008
					TON+NH_3	1973	2008
					NO_3	1950	2008
					Alk	1950	2008
					TP	1960	2008
					TDP	1977	2008
[2]127	Clarke, 1924	San Joaquin River near Lathrop, Calif.	37°50'29"	121°19'06"	NO_3	1907	1908
					Alk	1907	1908
128	STORET	RSAN112 San Joaquin River at Airport Way Bridge, Vernalis, Calif.	37°40'18"	121°23'02"	TN	1955	1977
					NH_3	1955	1977
					TON + NH_3	1955	1977
					NO_3	1942	1977
					Alk	1938	1990
					TP	1955	1977
129	STORET	WB05B0704000 San Joaquin River at Maze Bridge Road, Calif.	37°38'24"	121°13'37"	TN	1962	1968
					NH_3	1962	1978
					TON + NH_3	1962	1978
					NO_3	1934	1978
					Alk	1934	1978
					TP	1962	1978
130	STORET	WB05B0708000 San Joaquin River near Grayson, Calif.	37°33'48"	121°09'07"	TN	1961	1978
					NH_3	1961	1978
					TON+NH_3	1961	1978
					NO_3	1932	1978
					Alk	1932	1978
					TP	1962	1978

[1]Reference stream gaging station.

[2]Historical station, locatoin approximate.

Willamette River

The Willamette River is the largest river in Oregon, flowing north through the Willamette Valley for approximately 480 km (300 mi) to join the Columbia River at Portland, Oreg. (fig. 31). The sampling site referenced in this study is located in Salem, Oreg., near the center of the basin. The basin is located in the Pacific Border and Cascad-Sierra Mountains physiographic provinces (fig. 2). It is roughly rectangular in shape and drains about 18,900 km² (7,280 mi²) at the reference stream gaging station at Salem (station No. 14191000). The Coastal Mountains are the western boundary of the basin and the Cascades Mountains delimit the east and south (fig. 31). The Willamette Valley Plains are in the center of the river valley with the Willamette Valley Foothills surrounding (fig. 31) (Uhrich and Wentz, 1999). Major tributaries include the McKenzie River in the south, and the Santiam River in the central part of the basin.

The climate in this region of western Oregon is tempered by proximity to the Pacific Ocean, giving rise to temperate rain forest conditions that are characterized by two major seasons: relatively cool and wet winters, and warm and dry summers. With a mean annual discharge (1909–2009) at Salem of about 660 m³/s (23,000 ft³/s), runoff is about 1,100 mm/yr. Considering the basin as a whole, the Willamette River basin produces more runoff per unit of drainage area than any other large river in the conterminous United States (Kammerer, 1990). Precipitation falls primarily during October–March, as snow at high elevations and rainfall on the valley floor. The seasonal distribution of streamflow reflects the precipitation patterns, with about 60–85 percent of runoff occurring during the winter season (Uhrich and Wentz, 1999). Very low discharge during the summer months has been a major factor affecting water quality in the basin.

The most important land-use activity in the basin upstream of Salem is timber production, which became a major industry in the early 20th century. Effluent from the associated pulp and paper industry was a major water quality concern in the 1930s and 1940s (Gleeson, 1972). Additionally, because the river served historically as the primary transportation link within the valley, numerous settlements were established in close proximity to the main stem river, including the three largest cities in the State (Portland, Salem, and Eugene). Population density in the basin was 29 people/km² according to the year 2000 census (fig. 3). Because of the accessibility of the river to these towns and cities, one of the most important water-quality issues in the early 20th century was discharge of untreated sewage wastes to the river (Gleeson, 1972).

As a result of public concerns about pollution and coincident oxygen depletion from municipal and industrial wastes, a major effort to clean up the Willamette River occurred during the middle of the 20th century. This effort was focused primarily on improving treatment of wastewater and effluent from the paper industry (Gleeson, 1972). Another major component was construction of 13 flood control reservoirs by the U.S. Army Corps of Engineers in the upper reaches of the basin that allowed streamflow augmentation during the critical low-flow months in late summer. These reservoirs exert an important control on the flow in the Willamette River, with storage capacity equal to nearly 1.9 million acre-ft (Shearman, 1976). By 1972, the restoration of the Willamette River had become a national success story after achieving the desired goals for dissolved oxygen during low-flow conditions (Gleeson, 1972).

Agriculture has been an important component of land use on the valley floor since the first settlers arrived on the Oregon Trail in the mid-19th century. The fertility of the Willamette Valley soils was a major draw to early pioneers, who grew wheat, oats, potatoes and other vegetables, as well as fruit orchards; limited livestock raised included beef and dairy cattle. Major crops today include grass seed, Christmas trees, berries, filberts, and peppermint (Uhrich and Wentz, 1999). In 2002, 20 percent of the basin was in agricultural land usage (fig. 4). Nitrogen and phosphorus from fertilizer and animal livestock generally are low with fertilizer inputs being a larger proportion of these inputs (figs. 5A–5B). These sources were less than one-quarter of the largest mean annual sources in the corn belt of the Midwestern United States (figs. 5A–5B), and have been consistently low throughout the 20th century (figs. 28E–28F). Nonetheless, nitrogen fertilizer use in the Willamette River basin was an order of magnitude greater than in other western basins like the Colorado that generally are associated with application rates less than 0.2 g (N/m²)/yr (figs. 5A–5B) During 2000–2010, the corn harvest rate in the Willamette River basin was the lowest of any river basin considered in this study, 0.06 bushels/km² of drainage area (fig. 6).

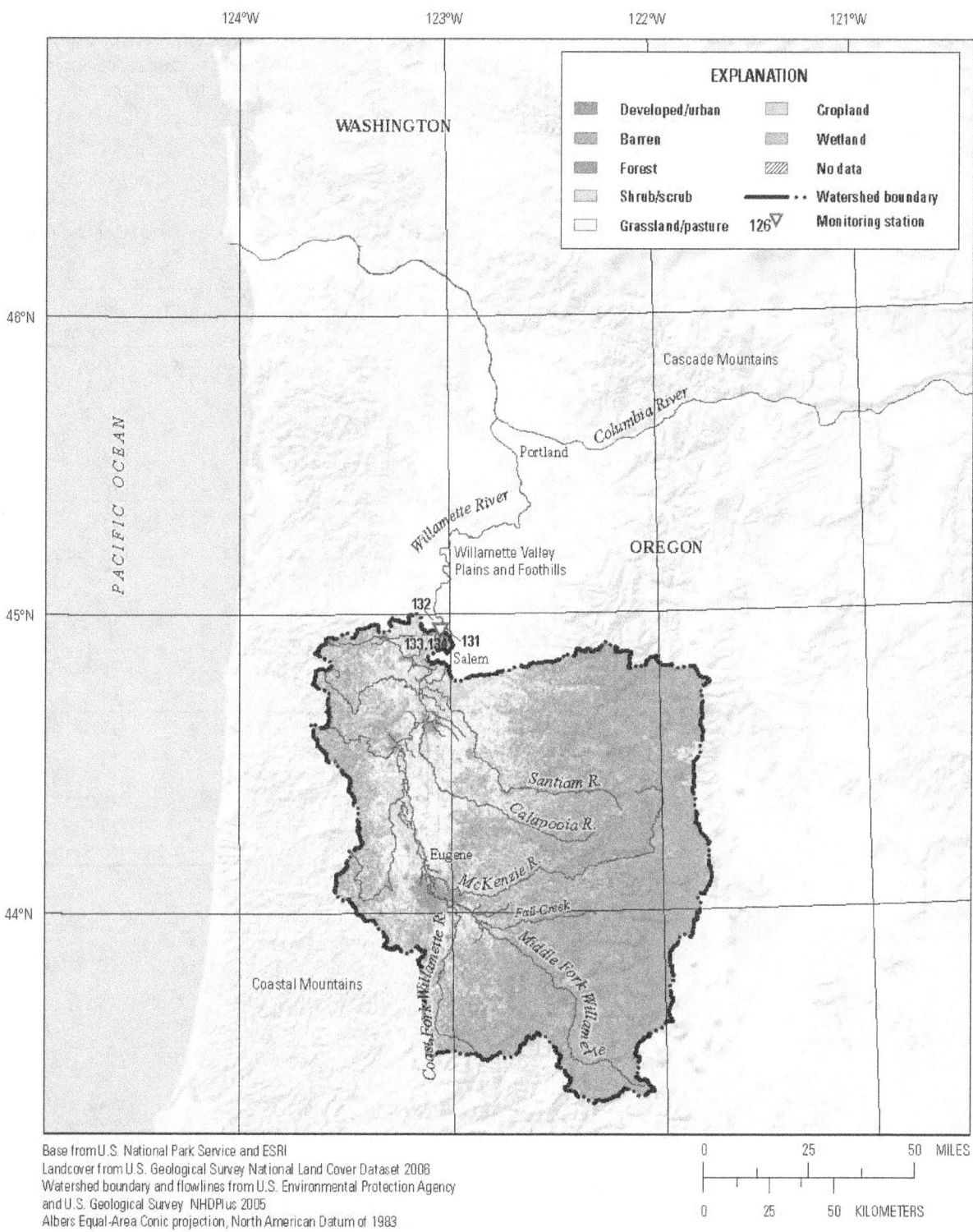

Figure 31. Willamette River basin.

Data Sources

Data sources for the Willamette River basin are summarized in table 21. Daily streamflow data have been collected by USGS at Salem, Oreg., since October 1909 (table 21). Water-quality data from 1911 to 1912 are available for the Willamette River at Salem from Clarke (1924), including major ions and nitrate. Additional data are available from the USGS and the Oregon Department of Environmental Quality, beginning in 1951 through the second half of the 20th century.

Table 21. Willamette River basin data sources.

[**Monitoring station** is shown in figure 31. **Source:** NWIS, National Water Information System; STORET, Storage and Retrieval Data Warehouse. **Constituent:** Alk, alkalinity. See table 2 for definitions of all other constituent abbreviations]

Monitoring station	Source	Station	Latitude	Longitude	Constituent	Start year	End year
[1]131	NWIS	14191000 Willamette River at Salem, Oreg.	44°56'40''	123°02'30''	Daily streamflow	1909	2009
					TN	1971	1972
					TON+NH$_3$	1971	1972
					NO$_3$	1951	1979
					Alk	1951	1979
					TDP	1970	1979
[2]132	Clarke, 1924	Willamette River at Salem, Oreg.	44°56'40''	123°02'30''	NO$_3$	1911	1912
					Alk	1911	1912
133	STORET	402014 Willamette River at Salem Railroad Bridge, Oreg.	44°56'52''	123°02'42''	TN	1982	1991
					NH$_3$	1959	1995
					TON+NH$_3$	1982	1991
					NO$_3$	1965	1995
					Alk	1959	1995
					TP	1965	1971
134	STORET	402226 Willamette River at Marion Street, Salem, Oreg.	44°56'46''	123°02'31''	TN	1991	1998
					NH$_3$	1991	1998
					TON+NH$_3$	1991	1998
					NO$_3$	1991	1998
					Alk	1991	1998
					TP	1991	1998

[1]Reference stream gaging station.

[2]Historical station, location approximate.

Summary

The issue of excess nutrients in streams and rivers of the Nation has been a critical one throughout the 20th century. Nutrient pollution has resulted in eutrophication of receiving waters and the coincident loss of biodiversity and development of hypoxia in coastal areas, especially the Gulf of Mexico. Adequately describing the origins and severity of nutrient pollution requires a large amount of data accumulated over a long period of time. In this study, we have provided evidence that long-term data exist for 26 drainage basins of interest including some of the largest rivers in the Nation. Utilizing these data will provide greater insight into the causes and trends in nutrient pollution.

The uneven geographic distribution of nutrient pollution over the past century is reflected in the unique history of water-quality issues in each of the drainage basins included in this study. On a national scale, water-quality problems can be considered to have begun in the middle 19th century as municipal water usage experienced its first wave of growth. An intense and largely successful effort to improve water sanitation for public health purposes occurred during the early and middle part of the 20th century. This effort focused on ensuring the safety of drinking water supplies, but in most areas wastewater was not treated. Even as water sanitation improved, other water-quality problems arose including increased pollution loads from industry, municipalities, and agriculture. In some cases, hydrologic development also created or exacerbated water-quality problems. Growth in population and affluence during the middle part of the 20th century also placed increasing pressures on water supplies and caused further water-quality degradation. Degraded in-stream conditions led to periods of anoxia, fish kills, and reduced utility of water resources.

Dramatic changes in agricultural practices occurred in the middle part of the 20th century and had important consequences for quality of water resources. Chemical fertilizer usage increased dramatically beginning in the late 1940s and led to tremendous increases in agricultural productivity, but has also been connected with increased nutrient pollution in rivers. In the 1950s, the usage of confined animal feeding operations (CAFOs) in livestock production became more common. By the end of the 20th century, most of the animals raised for human consumption are produced in CAFOs. Concentrating livestock population and concentrates livestock waste such that CAFOs can also be a source of nutrient pollution to aquatic ecosystems.

Local efforts to address water-quality concerns have been ongoing since the late 19th century. National-scale efforts developed more slowly with the first attempt at controlling water pollution nationwide being passed in the mid-20th century. The 1948 Water Pollution Control Act was a relatively ineffective law, but it laid the foundation for subsequent improvements, culminating in the passage of the Clean Water Act in 1972.

The Clean Water Act was passed with the stated goal of restoring and maintaining the integrity of the Nation's waters. This legislation was significant because it required dischargers to obtain a permit for their effluent, thereby defining a limit to the pollution load that could be discharged. As a result, treatment of effluent wastewater was greatly enhanced, especially focused on municipal and industrial point sources. Additionally, land-use planning was initiated to limit runoff from nonpoint sources in an attempt to solve the full range of pollution problems in the country. More recently, a range of research programs have been implemented to investigate the relation between land use and nonpoint pollution, evaluating soil conservation practices and runoff from farm fields and urban areas.

This study is focused on compiling data from selected river basins that will be useful to evaluate how these long-term changes in land use and policy have coincided with changes in nutrient concentrations. Sources for long-term streamflow and nutrient data are presented for 26 river basins, ranging across the continental United States. Brief histories are provided for each basin to describe the important events that are presumed to influence nutrient concentration patterns throughout the 20th century. Various issues of data assembly and management are discussed, including algorithms used to synthesize relevant data from disparate collecting agencies. Ancillary data documenting streamflow, population density, and selected land-use characteristics provide context for comparing these rivers with one another. The perspective provided by these data is unprecedented, although challenging to develop over such a lengthy period. Regardless, a long-term view of nutrient concentrations is essential for correctly understanding the impact of policies and legislation in controlling the delivery of excess nutrients to streams and rivers of the Nation.

References Cited

Adler, R.W., Landman, J.C., and Cameron, D.M., 1993, The Clean Water Act 20 Years Later: Island Press, Washington, D.C., 320 p.

Albert, R.C., 1982, Cleaning up the Delaware River: A status and progress report prepared under the auspices of Section 305(b) of the Federal Clean Water Act, Delaware River Basin Commission, West Trenton New Jersey. 46 p.

Alexander, R.B., and Smith, R.A., 1990, County-level estimates of nitrogen and phosphorus fertilizer use in the United States, 1945 to 1985: U.S. Geological Survey Open-File Report 90-130, 12 p.

American Water Resources Association, 1923, Report of industrial wastes in relation to water supply: Journal of American Water Resources Association, v. 10, p. 415-430.

Apodaca, L.E., Driver, N.E., Stephens, V.C., and Spahr, N.E., 1996, Environmental setting and implications on water quality, upper Colorado River Basin, Colorado and Utah: U.S. Geological Survey Water-Resources Investigations Report 95-4263, 33 p.

Bartow, E., 1907, Municipal water supplies of Illinois: University of Illinois Bulletin, Urbana, Ill., Water Survey Series No. 5, 129 p.

Bartow, E., 1909, Chemical and biological survey of the waters of Illinois: University of Illinois Bulletin, Urbana, Ill., Water Survey Series No. 7, 205 p.

Barnes, I., 1964, Field measurement of alkalinity and pH: U.S. Geological Survey Water-Supply Paper 1535-H, 17 p.

Barry, J.M., 1997, Rising tide—The great Mississippi flood of 1927 and how it changed America: New York, Simon & Schuster, 524 p.

Belval, D.L., Woodside, M.D., and Campbell, J.P., 1994, Relation of stream quality to streamflow, and estimated loads of selected water-quality constituents in the James and Rappahannock Rivers near the fall line of Virginia, July 1988 through June 1990: U.S. Geological Survey Water-Resources Investigations Report 94-4042, 85 p.

Benke, A.C., and Cushing, C.E., eds., 2005, Rivers of North America: Elsevier Academic Press, 1,144 p.

Bishop, A.B. and Porcella, D.B., 1980, Physical and ecological aspects of the upper Colorado River Basin, *in* Spofford, W.A., Parker, A.L., and Kneese, A.V., eds., Energy development in the southwest—Problems of water, fish and wildlife in the upper Colorado River basin: Washington D.C, Resources for the Future, v. 1, p. 17-56.

Bowers, D.E., Rasmussen, W.D., and Baker, G.L., 1985, History of agricultural price-support and adjustment programs, 1933-1984—Background for 1985 Farm Legislation: U.S. Department of Agriculture Agricultural Information Bulletin 485, 47 p.

Broussard, W., and Turner, R.E., 2009, A century of changing land-use and water-quality relationships in the continental U.S.: Frontiers in Ecology and the Environment, v. 7, p. 302-307.

Bureau of Reclamation, 2010, Upper Colorado River Basin consumptive uses and losses report 2006–2010: Bureau of Reclamation, November 2010, 29 p.

Burkholder, J, Libra, B., Weyer, P., Heathcote, S. Kolpin, D., Thorne, P.S., and Wichman, M., 2007, Impacts of waste from concentrated animal feeding operations on water quality: Environmental Health Perspectives, no. 115, p. 308-312.

Burton, C.A., Izbicki, J.A., and Paybins, K.S., 1998, Water-quality trends in the Santa Ana River at MWD Crossing and below Prado Dam, Riverside County, California: U.S. Geological Survey Water-Resources Investigations Report 97-4173, 36 p.

Butler, D.L., Krueger, R.P., Osmundson, B.C., Thompson, A.L., and McCall, S.K., 1991, Reconnaissance investigation of water quality, bottom sediment, and biota associated with irrigation drainage in the Gunnison and Uncompahgre River Basins and at Sweitzer Lake, west-central Colorado, 1988–89: U.S. Geological Survey Water-Resources Investigations Report 91-4103, 99 p.

California Regional Water Quality Control Board, 1990, Dairies and their relationship to water quality problems in the Chino Basin: California Regional Water Quality Control Board, 129 p.

California Regional Water Quality Control Board, 1995, Water Quality Control Plan—Santa Ana River Basin: California Regional Water Quality Control Board, 204 p.

Chesrow, C.F.W., and Hurwitz, E., 1961, Water pollution—A threat to Great Lakes cities, *in* Water Pollution and the Great Lakes: Chicago, Ill., DePaul University, 104 p.

City of Toledo, 2010, Plant history: City of Toledo, Web page, accessed February 17, 2012, at http://www.ci.toledo.oh.us/ Departments/PublicUtilities/DivisionofWaterReclamation/ PlantHistory/tabid/384/Default.aspx.

Clarke, F.W., 1924, The composition of the river and lake waters of the United States: U.S. Geological Survey Professional Paper 135, 215 p.

Cleary, E.J., 1967, The ORSANCO Story: water quality management in the Ohio valley under an interstate compact: Published for Resources for the Future by the Johns Hopkins Press, Baltimore, 335 p.

Cohen, B.R., 2009, Encyclopedia Virginia—Modern environmental history of Virginia: Virginia Foundation for the Humanities Web page, accessed February 17, 2012, at http://www.EncyclopediaVirginia.org/Modern_ Environmental_History_of_Virginia.

Collins, W.D. and C.S. Howard, 1928. Quality of the surface waters of New Jersey. U.S. Geological Survey Water Supply Paper 596-E. 266 p.

Criss, R.E., and Wilson, D.A., eds., 2003, At the Confluence—Rivers, floods, and water quality in the St. Louis region: St. Louis, Mo., Missouri Botanical Garden Press, 278 p.

Crohurst, W.R., 1933. A study of the pollution and natural purification of the Ohio River: IV. A resurvey of the Ohio River between Cincinnati, Ohio, and Louisville, Ky., including a discussion of the effects of canalization and changes in sanitary conditions since 1914-1916: U.S. Public Health Service, Public Health Bulletin No. 204, 111 p.

Cumming, H.S., Purdy, W.C., and Ritter, H.P., 1916, Investigation of the pollution and sanitary conditions of the Potomac Watershed, with special reference to self purification and the sanitary condition of shellfish in the Lower Potomac River: U.S. Public Health Service, Hygienic Laboratory Bulletin No. 104, 239 p.

Curtis, J.B., 2002, Fractured shale-gas systems: American Association of Petroleum Geologist Bulletin, v. 86, no.11, p. 1921-1938.

Davis, C.C., 1961, The biotic community in the Great Lakes with respect to pollution, *in* Conference on water pollution and the Great Lakes, May 15–1961, Proceedings: Chicago, Ill. DePaul University, p. 80-87 p.

De Paul University, 1961, Water pollution and the Great Lakes, *in* Conference to focus additional public attention on the problems and solutions involved in keeping our water supply usable, now and in the years ahead, May 15–16, 1961, Proceedings: Chicago Ill., De Paul University, 104 p.

Delaware River Basin Commission, 2011, Natural gas drilling in the Delaware River Basin: Delaware River Basin Commission, Web page accessed February 17, 2012, at http://www.nj.gov/drbc/programs/natural/.

Diaz, R.J., 2001, Overview of hypoxia around the world: Journal of Environmental Quality, v. 30, p. 275-281.

Edmondson, W.T., 1991, The uses of ecology: Seattle, Wash., University of Washington Press, 329 p.

Ellis, M.M., 1936, Erosion silt as a factor in aquatic environments: Ecology, v. 17, p. 29-42.

Ellis, M.M., 1943, A study of the Mississippi River from Chain of Rocks, St. Louis, Missouri, to Cairo, Illinois, with special reference to the proposed introduction of ground garbage into the river by the city of St. Louis: Chicago, Ill., U.S. Fish and Wildlife Service Special Scientific Report 8, 22 p.

Everts, C.M., Jr., and Dahl, A.H., 1957, The Federal Water Pollution Control Act of 1956: Journal of American Public Health, v. 47, p. 305-310.

Federal Water Pollution Control Administration, 1969, Potomac River water quality—Washington, D.C. metropolitan area: U.S. Department of the Interior, 80 p.

Fischer, J.M., Riva-Murray, K., Hickman, R.E., Chichester, D.C., Brightbill, F.A., Romanok, K.M., and Bilger, M.D., 2004, Water quality in the Delaware River basin: U.S. Geological Survey Circular 1227, 48 p.

Fuller, G.W., 1898, Report on the investigations into the purification of the Ohio River water at Louisville, Kentucky, made to the president and directors of the Louisville water company. Van Nostrand Co., New York. 300 p.

Fuller, K., Shear, H., and Wittig, J., 1995, The Great Lakes—An environmental atlas and resource book: Chicago, Ill., U.S. Environmental Protection Agency, 46 p.

Galloway, J.N., Dentener, F.J., Capone, D.G., Boyer, E.W., Howarth, R.W., Seitzinger, S.P., Asner, G.P., Cleveland, C.C., Green, P.A., Holland, E.A., Karl, D.M., Michaels, A.F., Porter, J.H., Townsend, A.R., and Vörösmarty, C.J., 2004, Nitrogen cycles—Past, present, and future: Biogeochemistry, no. 70, p. 153-226.

Garabedian, S.P., Coles, J.F., Grady, S.J., Trench, E.C.T., and Zimmerman, M.J., 1998, Water quality in the Connecticut, Housatonic, and Thames River Basins, Connecticut, Massachusetts, New Hampshire, New York, and Vermont, 1992–95: U.S. Geological Survey Circular 1155, 32 p.

Gleeson, G.W., 1972, The return of a river—The Willamette River, Oregon: Corvallis, Oreg. Oregon State University, 103 p.

Gloss, S.P., Reynolds, R.C., Jr., Mayer, L.M., add Kidd, D.E., 1981, Reservoir influences on salinity and nutrient fluxes in the arid Colorado River Basin, *in* H.G. Stefan, ed., Proceedings of the Symposium on Surface Water Impoundments: New York, American Society of Civil Engineers, p. 1618-1629.

Goolsby, D.A., Battaglin, W.A., Lawrence, G.B., Artz, R.S., Aulenbach, B.T., Hooper, R.P., Keeney, D.R., and Stensland, G.J., 1999, Flux and sources of nutrients in the Mississippi-Atchafalaya River basin—Topic 3, report for the integrated assessment on hypoxia in the Gulf of Mexico: National Oceanic and Atmospheric Administration, Coastal Ocean Program, Decision Analysis Series, no. 17, 129 p.

Grason, D., and Healy, R.W., 1979, Chemical analyses of surface water in Illinois, 1975–1977; Volume 2, Illinois River basin and Mississippi River tributaries north of Illinois River basin: U.S. Geological Survey Water-Resources Investigations 79-24, 287 p.

Great Lakes Fishery Commission, 2000, Sea Lamprey—A Great Lakes Invader: Great Lakes Fishery Commission Fact Sheet 3, 2 p.

Great Lakes Information Network, 2012, Great Lakes fish consumption advisories, Web Page accessed July 12, 2012 at: http://www.great-lakes.net/envt/flora-fauna/wildlife/fishadv.html#resources.

Gronberg, J.A.M., Dubrovsky, N.M., Kratzer, C.R., Domagalski, J.L., Brown, L.R., and Burow, K.R., 1998, Environmental setting of the San Joaquin-Tulare basins, California: U.S. Geological Survey Water-Resources Investigations Report 97-4205, 45 p.

Groschen, G.E., Harris, M.A., King, R.B., Terrio, P.J., and Warner, K.I., 2000, Water quality in the Lower Illinois River basin, Illinois, 1995–1998: U.S. Geological Survey Circular 1209, 36 p.

Haines, M.R., and Inter-university Consortium for Political and Social Research, 2004, Historical, demographic, economic, and social data—The United States, 1790–2000 [Computer file], ICPSR02896-v3: Ann Arbor, Mich., The Inter-university Consortium for Political and Social Research [distributor], accessed February 17, 2012, at http://dx.doi.org/10.3886/ICPSR02896.v3.

Hansen, Z.K., and Libecap, G.D., 2004, Small farms, externalities, and the Dust Bowl of the 1930s: Journal of Political Economy, no. 112, p. 665-694.

Harmeson, R.H., and T.E. Larson, 1969, Quality of surface water in Illinois, 1956-1966: Illinois State Water Survey, Urbana, Ill., Bulletin No. 54, 189 p.

Harmeson, R.H., Larson, T.E., Henley, L.M., Sinclair, R.A., and Neill, J.C., 1973, Quality of surface water in Illinois, 1966-1971: Illinois State Water Survey, Urbana, Ill., Bulletin No. 56, 104 p.

Harrington, S., and Harrington, J., 2009, The particulate fraction of nutrients transported in river loads in the South West River Basin District (SWRBD): Ireland, Geophysical research abstracts, v. 11, EGU2009-7684, 2 p.

Healy, R.W., and Toler, L.G., 1978, Chemical analyses of surface water in Illinois, 1958–1974—Volume II, Illinois River basin and Mississippi River tributaries: U.S. Geological Survey Water-Resources Investigations Report 78-23, 442 p.

Heusinkveld, H., 1989, Saga of the Des Moines River Greenbelt: Pella, Iowa, Pella Printing Company, 120 p.

Hickman, R.E., 2004, Pesticide compounds in streamwater in the Delaware River Basin, December 1998–August 2001: U.S. Geological Survey Scientific Investigations Report 2004-5105, 47 p.

Hill, L., 2000, The Chicago River—A natural and unnatural history, Chicago, Ill., Lake Claremont Press, 302 p.

Hoskins, J.K., Ruchhoft, C.C., Williams, L.G., and Purdy, W.C., 1927, A study of the pollution and natural purification of the Illinois River: Public Health Bulletin, no. 171, p. 1-208.

Hupfer, M.E., 1965, Forty years of water pollution control in Connecticut: Connecticut Water Resources Commission, 46 p.

Iowa Division of Public Health Engineering, 1934, Investigation of pollution on the Des Moines River: Iowa Department of Health, 118 p.

Izbicki, J.A., Mendez, G.O., and Burton, C.A., 2000, Stormflow chemistry in the Santa Ana River below Prado Dam and at the diversion downstream from Imperial Highway, Southern California, 1995-98: U.S. Geological Survey Water-Resources Investigations Report 00-4127, 92 p.

Jaworski, N.A., Howarth, R.W., and Hetling, L.J., 1997, Atmospheric deposition of nitrogen oxides onto the landscape contributes to coastal eutrophication in the northeast United States: Environmental Science and Technology no. 31, p. 1995-2004.

Jaworski, N.A., Romano, W., Buchanan, C., and Jaworski, C., 2007, The Potomac River basin and its estuary—Landscape loadings and water quality trends, 1895–2005: Interstate Commission on the Potomac River, Potomac Integrative Analysis Online Collection, p. 228, accessed February 17, 2012, at http://www.potomacriver.org/cms/index.php?option=com_content&view=article&id=147-pia-treatise&catid=37-assessing&Itemid=127.

Johnson, N.M., and Merritt, D.H., 1979, Convective and advective circulation of Lake Powell, Utah and Arizona, during 1972–1975: Water Resources Research, no. 15, p. 873-884.

Kammerer, J.C., 1990, Largest rivers in the United States: U.S. Geological Survey Open-File Report 87-242, 2 p.

Kargbo, D.M., Wilhelm, R.G., and Campbell, D.J., 2010, Natural gas plays in the Marcellus Shale—Challenges and potential opportunities: Environmental Science and Technology, no. 44, p. 5679-5684.

Keeney, D.R., and DeLuca, T.H., 1993, Des Moines River nitrate in relation to watershed agricultural practices—1945 versus 1980s: Journal of Environmental Quality, no. 22, p. 267-272.

Kirkpatrick, L.R., 2000, The San Juan River—The current controversy—Water, growth and sustainability—Planning for the 21st century, December 2000: New Mexico Water Resources Research Institute Report, p. 1-10 p.

Kiry, P.R., 1974, An historical look at the water quality of the Delaware River Estuary to 1973: Philadelphia, Pa., Academy of Natural Sciences of Philadelphia, Department of Limnology, 135 p.

Kratzer, C.R., and Shelton, J.L., 1998, Water quality assessment of the San Joaquin-Tulare basins, California—Analysis of available data on nutrients and suspended sediment in surface water, 1972–1990: U.S. Geological Survey Professional Paper 1587, 92 p.

Kratzer, C.R., Kent, R.H., Saleh, D.H., Knifong, D.L., Deleanis, P.D., and Orlando, J.L., 2011, Trends in nutrient concentrations, loads, and yields in streams in the Sacramento, San Joaquin, and Santa Ana Basins, California, 1975–2004: U.S. Geological Survey Scientific Investigations Report 2010-5228, 112 p.

Kusnetz, N., 2011, EPA plans to issue rules covering fracking wastewater: ProPublica, Oct. 20, 2011, accessed February 17, 2012, at http://www.propublica.org/article/epa-plans-to-issue-rules-covering-fracking-wastewater.

Le Quéré, C., Raupach, M.R., Canadell, J.G., Marland, G., Bopp, L., Ciais, P., Conway, T.J., Doney, S.C., Feely, R.A., Foster, P., Friedlingstein, P., Gurney, K., Houghton, R.A., House, J.I., Huntingford, C., Levy, P.E., Lomas, M.R., Majkut, J., Metzl, N., Ometto, J.P., Peters, G.P., Prentice, I.C., Randerson, J.T., Running, S.W., Sarmiento, J.L., Schuster, U., Sitch, S., Takahashi, T., Viovy, N., van der Werf, G.R., and Woodward, F.I., 2009, Trends in the sources and sinks of carbon dioxide: Nature Geoscience, v. 2, no. 12, p. 831-836.

Larson, T.E., and Larson, B.O., 1957, Quality of surface waters in Illinois, 1945-1955: State Water Survey Division, Urbana, Ill., Bulletin No. 45, 140 p.

Leighton, M.O., 1903, Normal and polluted waters in northeastern United States. U.S. Geological Survey Water Supply Paper 79, 192 p.

Leighton, M.O., 1907, Pollution of Illinois and Mississippi Rivers by Chicago sewage—A digest of the testimony taken in the case of the State of Missouri v. the State of Illinois and the Sanitary district of Chicago: U.S. Geological Survey Water-Supply and Irrigation Paper No. 194, 369 p., 2 pls.

Lewis, S.J., 1906, Quality of water in the upper Ohio River Basin and at Erie, Pa.: U.S. Geological Survey Water-Supply and Irrigation Paper 161, 114 p.

Leverin, H.A., 1947, Industrial waters of Canada, report on investigations, 1934 to 1943: Canada Bureau of Mines, 109 p.

Litke, D.W.,1999, Review of phosphorus control measures in the United States and their effects on water quality: U.S. Geological Survey Water-Resources Investigations Report 99-4007, 38 p.

Long, J.H., 1889, Chemical investigations of the water supplies of Illinois, 1888–1889, in Preliminary report to the Illinois State Board of Health—Water supplies of Illinois and the pollution of its streams: Springfield, Ill., H.W. Hokker, Printer and Binder, p. 1-25.

Love, S.K., 1954, Quality of surface waters of the United States, 1950—Parts 1-4, North Atlantic slope basins to St. Lawrence River Basin: U.S. Geological Survey Water Supply Paper 1186, 344 p.

Love, S.K., 1955. Quality of surface waters of the United States, 1951— Parts 1-4, North Atlantic slope basins to St. Lawrence River Basin: U.S. Geological Survey Water Supply Paper 1197, 385 p.

Love, S.K., 1956a. Quality of surface waters of the United States, 1952—Parts 1-4, North Atlantic slope basins to St. Lawrence River Basin: U.S. Geological Survey Water Supply Paper 1250, 380 p.

Love, S.K., 1956b, Quality of surface waters of the United States, 1952—Parts 7 and 8, lower Mississippi River Basin and western Gulf of Mexico basins: U.S. Geological Survey Water Supply Paper 1252, 486 p.

Love, S.K., 1959a, Quality of surface waters of the United States, 1954—Parts 7 and 8, lower Mississippi River Basin and western Gulf of Mexico basins: U.S. Geological Survey Water Supply Paper 1352, 501 p.

Love, S.K., 1959b, Quality of surface waters of the United States, 1955—Parts 7 and 8, lower Mississippi River Basin and western Gulf of Mexico basins: U.S. Geological Survey Water Supply Paper 1402, 539 p.

Love, S.K., 1960, Quality of surface waters of the United States, 1956—Parts 7 and 8, lower Mississippi River Basin and western Gulf of Mexico basins: U.S. Geological Survey Water Supply Paper 1452, 469 p.

Love, S.K., 1961, Quality of surface waters of the United States, 1957—Parts 7 and 8, lower Mississippi River Basin and western Gulf of Mexico basins: U.S. Geological Survey Water Supply Paper 1522, 494 p.

Love, S.K., 1963, Quality of surface waters of the United States, 1958—Parts 7 and 8, lower Mississippi River Basin and western Gulf of Mexico basins: U.S. Geological Survey Water Supply Paper 1573, 588 p.

Lutts, R.H., 2004, Manna from God—The American chestnut trade in southwestern Virginia: Environmental History, v. 9, no. 3, p. 497-525.

MacDonald, J.M., and McBride, W.D., 2009, The transformation of U.S. livestock agriculture: scale, efficiency, and risks: U.S. Department of Agriculture, Electronic Information Bulletin no. 43, January 2009, 40 p., accessed February 17, 2012, at http://www.ers.usda.gov/Publications/EIB43/EIB43.pdf.

Maryland Department of Natural Resources, 2007, Maryland tributary strategy Upper Potomac River basin summary report for 1985–2005 data, Upper Potomac Tributary Team, Chesapeake Bay Tributary Strategies: Maryland Department of Natural Resources, 82 p., accessed April 3, 2012, at http://www.dnr.state.md.us/bay/pdfs/UPRBasinSum8505FINAL07.pdf.

Meade, R.H., and Moody, J.A., 2010, Causes for the decline of suspended-sediment discharge in the Mississippi River system, 1940–2007: Hydrological Processes, no. 24, p. 35-49.

Melosi, M.V., 2000, The sanitary city—Urban infrastructure in America from colonial times to the present: Baltimore, Md., Johns Hopkins University Press, 578 p.

Merchant, C., 2002, The Columbia guide to American environmental history, New York, Columbia University Press, 400 p.

Milazzo, P.C., 2006, Unlikely environmentalists—Congress and clean water, 1945–1972: Lawrence, Kans., University Press of Kansas, 312 p.

Mills, J.E., and Davis, T.L., 2000, The recovery of the North Branch—1940 to 2000 and beyond: Frostburg, Md., Maryland Department of the Environment, Bureau of Mines, Frostburg State University, 11 p.

Minnesota Population Center, 2011, National Historical Geographic Information System University of Minnesota database, accessed February 17, 2012, at http://www.nhgis.org.

Mullaney, J.R., 2004, Summary of water quality trends in the Connecticut River, 1968–1998: American Fisheries Society Monograph 9, p. 273-286.

Murphy, E.F., 1961, Water purity, a study in legal control of natural resources, Madison, Wisc., University of Wisconsin Press, 212 p.

Myers, D.N., Mezker, K.D., and Davis, S., 2000, Status and trends in suspended-sediment discharges, soil erosion, and conservation tillage in the Maumee River Basin—Ohio, Michigan, and Indiana: U.S. Geological Survey Water-Resources Investigations Report 00-4091, 38 p.

National Agricultural Statistics Service, 2010, Data and statistics: National Agricultural Statistics Service database, accessed February 17, 2012, at http://www.nass.usda.gov/Data_and_Statistics/index.asp.

National Research Council, 2004, Confronting the nation's water problems—The role of research: Washington, D.C., National Academic Press, 310 p.

Novotny, V., 2003, Water quality—Diffuse pollution and watershed management: New York, John Wiley and Sons, Inc., 864 p.

Oblinger Childress, C.J., Foreman, W.T., Connor, B.F., and Maloney, T.J., 1999, New reporting procedures based on long-term method detection levels and some considerations for interpretations of water-quality data provided by the U.S. Geological Survey National Water Quality Laboratory: U.S. Geological Survey Open-File Report 99-193, 19 p.

Oh, N.H., and Raymond, P.A., 2006, Contribution of agricultural liming to riverine bicarbonate export and CO_2 sequestration in the Ohio River basin: Global Biogeochemical Cycles, v. 20, GB3012, 17 p.

Paavola, J., 2006, Interstate water pollution problems and elusive federal water pollution policy in the United States, 1900–1948: Environment and History, v. 12, p. 435-465.

Palmer, A.W., 1897, Chemical survey of the waters of Illinois: Urbana, Ill., Preliminary Report, Illinois State Water Survey, 98 p.

Palmer, A.W., 1903, Chemical survey of the waters of Illinois: Urbana, Ill., Report for the Years 1897-1902, 316 p.

Parker, H.N., Willis, B., Bolster, R.H., Ashe, W.W., and Marsh, M.C., 1907, The Potomac River Basin—geographic history—rainfall and stream flow—pollution, typhoid fever, and character of water—relation of soils and forest cover to quality and quantity of surface water—effect of industrial wastes on fishes: U.S. Geological Survey Water-Supply and Irrigation Paper No. 192, 364 p.

Patrick, R., 1992, Surface water quality—Have the laws been successful?: Princeton, N.J., Princeton University Press, 198 p.

Paulsen, C.G., 1950, Quality of surface waters of the United States, 1946: U.S. Geological Survey Water Supply Paper 1050, 486 p.

Paulsen, C.G., 1952a, Quality of surface waters of the United States, 1947: U.S. Geological Survey Water Supply Paper 1102, 651 p.

Paulsen, C.G., 1952b, Quality of surface waters of the United States, 1948—Parts 7-14: U.S. Geological Survey Water Supply Paper 1133, 373 p.

Paulsen, C.G., 1953a, Quality of surface waters of the United States, 1948—Parts 1-6: U.S. Geological Survey Water Supply Paper 1132, 515 p.

Paulsen, C.G., 1953b, Quality of surface waters of the United States, 1949—Parts 1-6: U.S. Geological Survey Water Supply Paper 1162, 662 p.

Paulsen, C.G., 1953c, Quality of surface waters of the United States, 1949—Parts 7-14: U.S. Geological Survey Water Supply Paper 1163, 504 p.

Phelps, E.B., 1914, Studies on the self-purification of streams: U.S. Public Health Service Public Health Reports, v. 29, no. 33, p. 2128-2132.

Plummer, L.N., and Busenberg, E., 1982, The solubilities of calcite, aragonite and vaterite in CO_2-H_2O solutions between 0 and 90°C, and an evaluation of the aqueous model for the system $CaCO_3$-CO_2-H_2O: Geochimica et Cosmochimica Acta, v. 46, p. 1101-1040.

Pollan, M., 2006, The omnivore's dilemma—A natural history of four meals, New York, Penguin Press, 450 p.

Poston, H.W., 1961, What is water pollution, *in* Water Pollution and the Great Lakes: Chicago, Ill., DePaul University, 104 p.

Pudup, M.B., 1990, The limits of subsistence—Agriculture and industry in central Appalachia: Agricultural History, v. 64, no. 1, p. 61-89.

Purdy, W.C., 1923, A study of the pollution and natural purification of the Ohio River—I. The plankton and related organisms: U.S. Public Health Service Public Health Reports 131, 78 p.

Purdy, W.C., 1930, A study of the pollution and natural purification of the Illinois River—II. The plankton and related organisms: Public Health Bulletin 198, 212 p.

Rauch, J.H., 1889, Preliminary report to the Illinois State Board of Health—Water supplies of Illinois and the pollution of its streams: Springfield Ill., H.W. Hokker, Printer and Binder, 123 p.

Raymond, P., Oh, N.-H., Turner, R.E., and Broussard, W., 2008, Anthropogenically enhanced fluxes of water and carbon from the Mississippi River: Nature, v. 451, p. 449-452.

Robinson, K.W., Campbell, J.P., and Jaworski, N.A., 2003, Water-quality trends in New England Rivers during the 20th century: U.S. Geological Survey Water-Resources Investigations Report 03-4012, 20 p.

Ruddy, B.C., Lorenz, D.L., and Mueller, D.K., 2006. County-level estimates of nutrient input to the land-surface of the conterminous United States, 1982–2001: U.S. Geological Survey Scientific Investigations Report 2006-5012, 17 p.

Scarpino, P.V., 1985, Great River—An environmental history of the upper Mississippi, 1890–1950: Columbia, Mo., University of Missouri Press, 242 p.

Schaefer, S.C., and Alber, M., 2007, Temperature controls a latitudinal gradient in the proportion of watershed nitrogen exported to coastal ecosystems: Biogeochemistry, no. 85, p. 333-346.

Schindler, D.W., 1977, Evolution of phosphorus limitation in lakes: Science, v. 195, p. 260-262.

Schindler, D.W., 2006, Recent advances in the understanding and management of eutrophication: Limnology and Oceanography, v. 51, p. 356-363.

Schuylkill Watershed Conservation Plan, 2001, Schuylkill Watershed Conservation Plan:

Prepared for Pennsylvania Department of Conservation and Natural Resources and The William Penn Foundation, accessed February 17, 2012, at http://www.schuylkillplan.org/.

Seaber, P.R., Kapinos, F.P., and Knapp, G.L., 1987, Hydrologic unit maps: U.S. Geological Survey Water-Supply Paper 2294, 63 p.

Sharpley, A. and Moyer, B., 2000, Phosphorus forms in manure and compost and their release during simulate rainfall: Journal of Environmental Quality, v. 29, p. 1462-1469.

Shearman, J.O., 1976, Reservoir-system model for the Willamette River Basin, Oregon: U.S. Geological Survey Circular 715-H, 22 p.

Sprague, L.A., Clark, M.L., Rus, D.L., Zelt, R.B., Flynn, J.L., and Davis, J.V., 2007, Nutrient and suspended-sediment trends in the Missouri River Basin, 1993–2003: U.S. Geological Survey Scientific Investigations Report 06-5231, 80 p.

Sprague, L.A., Langland, M.J., Yochum, S.E., Edwards, R.E., Blomquist, J.D., Phillips, S.W., Shenk, G.W., and Preston, S.D., 2000, Factors affecting nutrient trends in major rivers of the Chesapeake Bay Watershed: U.S. Geological Survey Water-Resources Investigations Report 00-4218, 109 p.

Sprague, L.A., Mueller, D.K., Schwarz, G.E., and Lorenz, D.L., 2009, Nutrient trends in streams and rivers of the United States, 1993–2003: U.S. Geological Survey Scientific Investigations Report 2008-5202, 196 p.

State Board of Health of Indiana, 1911, A sanitary survey of the Ohio River: Thirtieth Annual Report of the State Board of Health of Indiana, 653 p.

Stamer, J.K., Yorke, T.H., and Perderson, G.L., 1985, Distribution and transport of trace substances in the Schuylkill River Basin from Berne to Philadelphia, Pennsylvania: U.S. Geological Survey Water Supply Paper 2256-A, 45 p.

Stanford, J.A., and Ward, J.V., 1991, Limnology of Lake Powell and the chemistry of the Colorado River, *in* Marzolf, G.R., ed., Colorado River ecology and dam management: Proceedings—Symposium, May 24–25, 1990, Santa Fe, New Mexico: Washington, D.C., National Academy Press, p. 75–101.

Streeter, H.W., and Phelps, E.B., 1925, A study of the pollution and natural purification of the Ohio River—III, Factors concerned in the phenomena of oxidation and reparation: Public Health Bulletin, v. 146, 75 p.

Tarr, J.A., 1996, The search for the ultimate sink: urban pollution in historical perspective: Akron, Ohio, University of Akron Press, 424 p.

Tetra Tech, 2000, Progress in water quality—An evaluation of the national investment in municipal wastewater treatment, Connecticut River Case Study:, U.S. Environmental Protection Agency, EPA-832-R-00-008, Chapter 5, 452 p.

Thompson, J., 2002, Wetlands drainage, river modification, and sectoral conflict in the lower Illinois Valley, 1890–1930: Carbondale, Ill., Southern Illinois University Press, 304 p.

Trench, E.C.T., 2000, Nutrient sources and loads in the Connecticut, Housatonic, and Thames River Basins: U.S. Geological Survey Water-Resources Investigations Report 99-4236, 66 p.

Twin Cities Metropolitan Council, 2006, 100+ years of water quality improvements in the Twin Cities—A chronology of significant events affecting water quality in the Mississippi River in the Twin Cities metropolitan area: St. Paul, Minn., Metropolitan Council Environmental Services, Fact sheet, 4 p., accessed April 3, 2012, at https://www.metrocouncil.org/environment/AboutMCES/.

U.S. Census Bureau, (2000), American FactFinder: U.S. Census Bureau database, accessed February 17, 2012, at http://factfinder2.census.gov/.

U.S. Department of Agriculture, 2005, Census of agriculture, 2002: U.S. Department of Agriculture database, accessed May 8, 2012, at http://www.agcensus.usda.gov/Publications/2002/index.php.

U.S. Department of the Interior, 2003, Quality of water—Colorado River Basin: U.S. Department of the Interior, Progress report no. 21, 95 p.

U.S. Department of the Interior, 1987, National register of historic places, farms in Berks County, Pa.: National Park Service, 55 p., accessed February 17, 2012 at http://www.phmc.state.pa.us/Portal/Communities/BHP/MPDFs/Farms_in_Berks_County_PA.pdf.

U.S. Department of the Interior, 1968, The nation's river: A report on the Potomac from the U.S. Department of the Interior, with recommendations for action by the Federal Interdepartmental Task Force on the Potomac, 112 p.

U.S. Environmental Protection Agency, 2006, Maumee area of concern—Stage 2 Watershed Restoration Plan, Volume 1 (draft), 232 p., accessed February 17, 2012, at http://www.epa.gov/glnpo/aoc/maumee/Maumee-AOC-Stage2Plan.pdf.

U.S. Environmental Protection Agency Science Advisory Board, 2007, Hypoxia in the Northern Gulf of Mexico—An update by the EPA Science Advisory Board: U.S. Environmental Protection Agency, EPA-SAB-08-003, 333 p.

U.S. Geological Survey, 1954, Quality of surface waters of the United States, 1950. Parts 1-4, North Atlantic slope basins to St. Lawrence River basin: U.S. Geological Survey Water Supply Paper 1186, 331 p.

U.S. Geological Survey, 1955, Quality of surface waters of the United States, 1951—Parts 1-4, North Atlantic slope basins to St. Lawrence basin: U.S. Geological Survey Water Supply Paper 1197, 373 p.

U.S. Geological Survey, 1955, Quality of surface waters of the United States, 1951—Parts 7 and 8, lower Mississippi River Basin and western Gulf of Mexico basins: U.S. Geological Survey Water Supply Paper 1199, 490 p.

U.S. Geological Survey, 1956, Quality of surface waters of the United States, 1952—Parts 1-4, North Atlantic slope basins to St. Lawrence basin: U.S. Geological Survey Water Supply Paper 1250, 365 p.

U.S. Geological Survey, 1958, Quality of surface waters of the United States, 1953—Parts 7 and 8, lower Mississippi River Basin and western Gulf of Mexico basins: U.S. Geological Survey Water Supply Paper 1292, 524 p.

U.S. Geological Survey, 1993, National water summary 1990–1991—Hydrologic events and stream water quality: U.S. Geological Survey Water Supply Paper 2400, 590 p.

U.S. Geological Survey, 1988, Quality of water branch technical memorandum 88.05: U.S. Geological Survey, accessed February 17, 2012 at http://pubs.usgs.gov/dds/wqn96cd/html/wqn/qasure/qw88_05.htm.

U.S. Geological Survey, 1998, Lower Missouri River Ecosystem Initiative Final Report 1994-1998: U.S. Geological Survey Columbia Environmental Research Center, 20 p., accessed February 17, 2012, http://infolink.cr.usgs.gov/AboutInfoLINK/lmreifinal.pdf.

U.S. Geological Survey, 2011, Zebra mussels cause economic and ecological problems in the Great Lakes: U.S. Geological Survey Great Lakes Science Center, Fact Sheet 2000-6, accessed July 12, 2012, http://www.glsc.usgs.gov/files/factsheets/2000-6%20Zebra%20Mussels.pdf.

U.S. Public Health Service, 1942, Final report to the Ohio River Committee: Ohio River Pollution Survey, 205 p.

Uhrich, M.A., and Wentz, D.A., 1999, Environmental setting of the Willamette Basin, Oregon: U.S. Geological Survey Water-Resources Investigations Report 97-4082-A, 20 p.

University of Wisconsin Sea Grant, 2002, Fish of the Great Lakes, Alewife: University of Wisconsin Sea Grant Institute Web site, accessed February 17, 2012, at http://seagrant.wisc.edu/greatlakesfish/alewife.html.

Virginia Department of Environmental Quality, 2005, Chesapeake Bay nutrient and sediment reduction tributary strategy for the James River, Lynnhaven and Poqhoson coastal basins, March 2005: Richmond, Va., Commonwealth of Virginia, 146 p.

Vogl, A.L., and Lopes, V.L., 2009, Impacts of water resources development on flow regimes in the Brazos River: Environmental Monitoring and Assessment, v. 157, p. 331-345.

Weston, R.S., 1903, Water purification investigation: Report on Water Purification Investigation and on Plans Proposed for Sewerage and Water-works Systems, Sewerage and Water Board, New Orleans, La., 346 p.

Wiebe, A.H., 1931, Dissolved phosphorus and inorganic nitrogen in the water of the Mississippi River: Science, v. 73, p. 652.

Williams, C.A., Gerner, S.J., and Elliot, J.G., 2009, Summary of fluvial sediment collected at selected sites on the Gunnison River in Colorado and the Green and Duchesne Rivers in Utah, water years 2005–2008: U.S. Geological Survey Data Series 409, 123 p. (Also available at http://pubs.er.usgs.gov/publication/ds409.)

Work Projects Administration, 1940. Report on sources of pollution: Millers River Valley, Masssachusetts. Boston, WPA Pollution Studies Project. 141 p.

Wurbs, R.A., Karama, A.S., Saleh, I., and Ganze, C.K., 1993, Natural salt pollution and water supply reliability in the Brazos River Basin: College Station, Tex., Texas A&M University, Texas Water Resources Institute, Technical Report No. 160, 177 p.

Zeng, F.W., Masiello, C.A., and Hockaday, W.C., 2011, Controls on the origin and cycling of riverine dissolved inorganic carbon in the Brazos River, Texas: Biogeochemistry, no. 104, p. 275-291.